SpringerBriefs in Computer Science

Series Editors
Stan Zdonik
Peng Ning
Shashi Shekhar
Jonathan Katz
Xindong Wu
Lakhmi C. Jain
David Padua
Xuemin Shen
Borko Furht
V.S. Subrahmanian
Martial Hebert
Katsushi Ikeuchi
Bruno Siciliano

W0192984

For further volumes:
http://www.springer.com/series/10028

Ping Wang • Weihua Zhuang

Distributed Medium Access Control in Wireless Networks

 Springer

Ping Wang
Nanyang Technological University
Singapore

Weihua Zhuang
University of Waterloo
Waterloo, ON, Canada

ISSN 2191-5768   ISSN 2191-5776 (electronic)
ISBN 978-1-4614-6601-7 ISBN 978-1-4614-6602-4 (eBook)
DOI 10.1007/978-1-4614-6602-4
Springer New York Heidelberg Dordrecht London

Library of Congress Control Number: 2013933061

Printed on acid-free paper

Springer is part of Springer Science+Business Media (www.springer.com)

Preface

Due to the hostile transmission environment and limited radio resources, quality-of-service (QoS) provisioning in wireless networks is much more complex and difficult than in its wired counterpart. Moreover, due to the lack of central controller in the networks, distributed network control is required, adding complexity to QoS provisioning. In this book, distributed medium access control (MAC) with QoS provisioning is investigated for both single- and multi-hop wireless networks including wireless local area networks (WLANs), wireless ad hoc networks, and wireless mesh networks. For WLANs, an efficient MAC scheme and a call admission control algorithm are proposed to provide guaranteed QoS for voice traffic and, at the same time, increase the voice capacity significantly compared with the current WLAN standard. In addition, a novel token-based scheduling scheme is proposed to provide great flexibility and facility to the network service provider for service class management. As a WLAN has small coverage and cannot meet the growing demand for wireless service requiring communications "at anywhere and at anytime," a large-scale multi-hop wireless network (e.g., wireless ad hoc networks and wireless mesh networks) becomes a necessity. Due to the location-dependent contentions, a number of problems (e.g., hidden/exposed terminal problem, unfairness, and priority reversal problem) appear in a multi-hop wireless environment, posing more challenges for QoS provisioning. To address these challenges, a novel busy-tone-based distributed MAC scheme for wireless ad hoc networks and a collision-free MAC scheme for wireless mesh networks are proposed, respectively, taking the different network characteristics into consideration. The proposed schemes enhance the QoS provisioning capability to real-time traffic and, at the same time, significantly improve the system throughput and fairness performance for data traffic, as compared with the most popular IEEE 802.11 MAC scheme.

Singapore   Ping Wang
Waterloo, ON, Canada   Weihua Zhuang

Contents

Acronyms

AC	Access category
ACK	Acknowledgement
AIFS	Arbitration interframe space
AP	Access point
ARQ	Automatic repeat request
BER	Bit error rate
CAC	Call admission control
CDMA	Code-division multiple access
CDF	Cumulative distribution function
CFP	Contention-free period
CP	Contention period
CSMA	Carrier sense multiple access
CSMA/CA	CSMA with collision avoidance
CSMA/CD	CSMA with collision detection
CTS	Clear to send
CW	Contention window
DBTMA	Dual busy tone multiple access
DCF	Distributed coordination function
DIFS	Distributed interframe space
DSP	Digital signal processing
DSSS	Direct sequence spreading spectrum
EDCA	Enhanced distributed channel access
GPS	Global positioning system
GSM	The global system for mobile communications
IP	Internet protocol
ISM	Industrial, scientific, and medical frequency band
MAC	Medium access control
MACA	Multiple access collision avoidance
MACAW	MACA for wireless LANs
MIMO	Multiple input multiple output
MS	Mobile station

OFDM	Orthogonal frequency-division multiplexing
OFDMA	Orthogonal frequency-division multiple access
PCF	Point coordination function
PDF	Probability density function
PMF	Probability mess function
PN	Pseudo noise
QoS	Quality-of-service
RTP	Real-time transport protocol
RTS	Request to send
SIFS	Short interframe space
TCP	Transmission control protocol
TDD	Time-division duplex
TDMA	Time-division multiple access
TXOP	Transmission opportunity
UDP	User datagram protocol
VoIP	Voice over IP
WiMAX	Worldwide interoperability for microwave access
WLAN	Wireless local area network

Chapter 1
Introduction

1.1 Heterogeneous Wireless Communication Networks

In the past decade, with the advances of wireless technologies and the increasing demand for wireless communication services, a variety of wireless networks have been deployed, e.g., cellular networks, wireless local area networks (WLANs), wireless ad hoc networks, sensor networks, and wireless mesh networks, etc. Among them, cellular networks and WLANs are the two most popular ones. Cellular networks can provide high-quality voice service with wide-area coverage and seamless roaming. Cellular networks evolve from the first generation (1G) based on analog technology, the second generation (2G) based on the digital technology, to the current third generation (3G) based on wideband code-division multiple access (CDMA) technology. The current 3G system supports a data rate up to 2 Mbps. The next generation (4G) has attracted much attention from academia, which is expected to provide multimedia services with much higher data rate [38].

As another popular wireless network, the WLAN has also achieved great success because of its simplicity, flexibility, high-rate access, and low cost. WLANs typically cover a small geographic area, in hot-spot local areas such as airports, malls, offices, and hotels, etc. The current WLAN standards are IEEE 802.11 series [1]. The IEEE 802.11b operates at the license-exempt 2.4 GHz industrial, scientific, and medical (ISM) frequency band, supporting a data rate up to 11 Mbps. The subsequent revisions 802.11a and 802.11g provide up to 54 Mbps data rate at the unlicensed 5 and 2.4 GHz bands, respectively, by employing orthogonal frequency-division multiplexing (OFDM) technology [57]. IEEE 802.11n, as the next generation WLAN standard, is expected to provide data rate as high as 200 Mbps by using multiple input multiple output (MIMO) technology [94], and have a higher market share in the next few years [96]. Cellular networks and WLANs are usually single-hop networks, where mobile users communicate with base stations (in cellular networks) or access point (in WLANs) via a direct wireless link.

P. Wang and W. Zhuang, *Distributed Medium Access Control in Wireless Networks*,
SpringerBriefs in Computer Science, DOI 10.1007/978-1-4614-6602-4_1,
© The Author(s) 2013

Different from the above two networks, a wireless ad hoc network is usually a multi-hop network, temporarily set up without any pre-existing infrastructure. Every node in such a network is functionally identical and may act as an end host and a router. Compared with cellular networks, an ad hoc network is a distributed network without central entities (i.e., base stations in cellular systems) for network organization and control, and the network can be set up on demand in a more timely manner with lower cost. Compared with WLANs, an ad hoc network usually has a larger coverage. These features make ad hoc networks well suited for situations where communication network infrastructures are either unavailable or difficult to set up, such as battle fields and disaster relief areas. Other attractive applications of ad hoc networks include temporary conference networks and home networks [79].

With the rapid growth of the Internet, there is an increasing demand for wireless broadband Internet access from both mobile and stationary users, using a less expensive and easier to deployment infrastructure than the wireline counterparts (such as digital subscriber line and cable). Wireless mesh networking is a promising wireless technology for future broadband Internet access, and has been attracting significant attention from both academia and industry [14]. It consists of wireline gateways, wireless routers, and mobile stations (MSs) [7]. Mesh routers are usually located at fixed sites and form a mesh backbone for MSs. As the routers establish and maintain mesh connectivity among themselves without a central controller, a wireless mesh network is generally considered as a type of ad hoc networks. Different from the traditional ad hoc networks, where the network topology may dynamically change due to the node mobility, a wireless mesh backbone usually has a static topology, and a mesh router can know the exact locations of other mesh routers. This feature can help to reduce the complexity of routing and medium access control (MAC) protocol design.

1.2   Quality-of-Service Provisioning in Wireless Networks

In recent years, with the rapid growth of Internet, there is an increasing popularity of multimedia applications. Typical applications include voice over IP (VoIP), video streaming, video conference, web browsing, and file transfer. With the integration of Internet and heterogeneous wireless networks, wireless networks are expected to ensure quality-of-service (QoS) for multimedia applications. QoS refers to a set of service requirements of selected traffic to be met by the network [21]. Different applications have different QoS requirements. For instance, real-time applications (such as voice and video) are delay sensitive but can tolerate some packet loss, while non-real-time applications (such as data applications) are delay insensitive but can tolerate little packet loss. The primary goal of QoS provisioning is to meet the different QoS requirements of users; meanwhile, from the service providers' point of view, the network resources should be efficiently utilized.

The unique characteristics of wireless networks make QoS provisioning a very complex and challenging task. The absence of a central controller, limited bandwidth, error-prone wireless channel, limited power, and the mobility of nodes impose many difficulties in providing QoS in such networks. The QoS provisioning can be achieved at different layers of the network protocol stack. Examples include multiple antennas at the physical layer, QoS-oriented scheduling at the MAC layer, QoS-aware routing at the network layer, and application adaptation at the application layer [59]. In this book, our focus is on the MAC layer and we assume that a QoS-aware routing protocol is available to choose a proper path from the source to the destination, and each node along a path is aware of the routing information.

The function of MAC is to coordinate the nodes in a network and to resolve the contention among their accessing the shared medium (i.e., the wireless channel) so that the limited radio resources are shared fairly and efficiently. Nodes in wireless networks usually operate in half-duplex mode and cannot transmit and receive simultaneously due to the fact that when a node's transmitter is transmitting, a large fraction of energy will leak into its receiving path, preventing the node from correct reception. As a result, collision detection is almost impossible and carrier sense multiple access with collision detection (CSMA/CD) cannot be deployed in wireless networks. Many wireless MAC schemes are based on carrier sense multiple access with collision avoidance (CSMA/CA) [79]. However, CSMA/CA does not provide any QoS provisioning feature. A QoS-oriented MAC scheme is required to provide prioritized access to let real-time traffic be transmitted in preference of data traffic, meanwhile achieve fairness (in terms of throughput) among data traffic. Since the wireless channel bandwidth is scarce, the QoS-oriented MAC scheme must achieve efficient channel utilization. With a given physical layer, a properly designed MAC scheme is the key to desired system performance such as fairness and high throughput to data traffic and short delay to real-time traffic.

1.3 The Importance and Challenges of MAC in Wireless Networks

Cellular networks are originally designed to provide high-quality voice service. The centralized control and reservation-based resource allocation enable fine QoS provisioning in cellular networks. In contrast, WLANs are originally designed for best-effort data applications without QoS assurance. Although many QoS enhancement mechanisms are proposed for WLANs [104], QoS provisioning capability of WLANs is still very limited in comparison with that of cellular networks. The system capacity for voice users is quite low in current WLANs [91]. Voice traffic may be interfered by other traffic (e.g., data traffic), resulting in a delay bound violation or large delay variance [72].

Although the current WLAN standard IEEE 802.11 series can provide a certain degree of service differentiation, it is difficult to quantify the degree of service differentiation, and even more difficult to adjust the degree flexibly among different classes based on some specific requirements of customers or network service providers. For example, when customers are charged differently for different services, it is desired that the received services (or resources) are proportional to what they are charged. Such kind of service model is referred to as proportional differentiation model [23], which assures that the performance of a class is proportional to that of another class according to a ratio preset by the network service provider. Such a feature provides great flexibility and facility to network service providers for service management. Most of the existing MAC schemes for WLANs (including the standard and its enhancements) are contention window based schemes without support for the proportional service differentiation.

QoS provisioning is relatively easy to be achieved in a centralized network (e.g., a cellular network) since the central controller (e.g., a base station) has sufficient information of the contending nodes, large processing power, and an efficient and collision-free way to broadcast the scheduling result to all the contending nodes. However, many wireless networks are distributed networks without a central controller (e.g., wireless ad hoc networks and wireless mesh networks). In such a network, each node does not have explicit information about other contenders, and there is no efficient way to let one node control the behavior of others. Hence, it is difficult to coordinate the transmissions from nodes in a distributed manner. This challenge adds more complexity to QoS provisioning and leads to intensive research work recently [1, 3, 6, 8, 63, 64, 68, 74, 80]. However, so far most of the work just focuses on single-hop networks, assuming that all the contending nodes can hear the transmission of each other. When applied to multi-hop networks, they may not work well because a multi-hop environment[1] presents more challenges to implement distributed MAC schemes than a single-hop environment. The existence of hidden terminals and exposed terminals (to be explained in Chap. 2) bring much more collisions and inefficient frequency reuse, respectively, leading to a significant degradation on the system throughput [33]. How to completely avoid the collisions due to hidden terminals in multi-hop networks is an open problem. Furthermore, the locations of the contending flows may heavily affect the channel access opportunity of each flow, resulting in serious unfairness (starvation of some flows) and priority reversal problems (i.e., a high-priority flow gets a smaller chance to access the channel than its low-priority counterpart) [99]. Although some research work has been done to address some of these problems [33, 39, 45, 56, 58], to the best of our knowledge, so far there is no comprehensive solution to address all the problems associated with a multi-hop environment. Without solving all these problems, QoS provisioning for multimedia applications is difficult to achieve.

This book presents several novel, effective and efficient QoS-oriented MAC schemes for multimedia traffic in heterogeneous single- and multi-hop wireless networks to address the above limitations. A WLAN is selected as a typical single-hop network, while wireless ad hoc networks and wireless mesh networks are

---

[1]In this book, we mostly focus on a multi-hop environment (i.e., a non-fully-connected network environment) but not a multi-hop flow.

selected as typical multi-hop networks. Specifically, we will present an efficient MAC scheme to significantly increase the system capacity for voice traffic in the current WLANs in Chap. 3. A theoretical model is developed to obtain the voice service capacity so that call admission control (CAC) can be facilitated to maximize the traffic load and guarantee QoS of voice traffic [83, 84]. A novel token-based MAC scheme will be presented in Chap. 4 to achieve proportional service differentiation in WLANs so that service classes can be flexibly adjusted based on specific requirements of customers, providing great flexibility and facility to the network service provider for service class management [88, 90]. In Chap. 5, we will introduce a distributed MAC with QoS provisioning for wireless ad hoc networks, and provide a comprehensive solution to address the hidden terminal, exposed terminal, priority reversal, and unfairness problems associated with the multi-hop network environment [82, 85, 87]. In Chap. 6, we will introduce a distributed collision-free MAC scheme to achieve high resource utilization and end-to-end QoS support for multimedia applications in a wireless mesh backbone [86, 89]. Different from the existing MAC schemes, our MAC scheme design benefits greatly from the fixed network topology of a wireless mesh backbone. With the router location information, collision-free transmissions can be scheduled, and the overhead is greatly reduced, as compared with conventional contention-based MAC schemes.

Chapter 2
Literature Review and Background

Heterogeneous wireless networks including single- and multi-hop wireless networks are considered. As WLANs are one of the most successful single-hop wireless network and has been widely deployed all over the world, we firstly study MAC layer QoS provisioning in WLANs. To meet the growing demand for wireless service requiring communications "at anywhere and at anytime", a large-scale multi-hop wireless network becomes a necessity. A multi-hop network environment presents more challenges to QoS supporting than a single-hop environment. In this chapter, we discuss these challenges in details and review the related research work.

2.1 MAC in WLANs

2.1.1 IEEE 802.11 MAC Protocol

As the WLAN standard, IEEE 802.11 [1] defines a mandatory distributed coordination function (DCF) and an optional centralized point coordination function (PCF). DCF is based on CSMA/CA. Each node randomly chooses a backoff timer from its contention window (CW). Before initiating a packet transmission, each contending node first senses the channel. After sensing the channel being idle for an 'distributed interframe space (DIFS)' duration, each node begins to count down its backoff timer after every idle slot until the backoff timer is decremented to zero, then the node starts to transmit, as shown in Fig. 2.1. If the channel is sensed busy before the timer goes to zero, the node freezes the timer and waits for the channel to be idle for another DIFS duration again, then continues to count down the timer. A positive acknowledgment (ACK) is used to notify the sender that the transmitted packet has been received successfully. If no ACK is received, the sender will schedule a retransmission. The CW is initially set to the minimum value CW_{min} for the first transmission attempt, doubled after each unsuccessful transmission until the maximum value CW_{max} is reached, and reset to CW_{min} after

P. Wang and W. Zhuang, *Distributed Medium Access Control in Wireless Networks*,
SpringerBriefs in Computer Science, DOI 10.1007/978-1-4614-6602-4_2,
© The Author(s) 2013

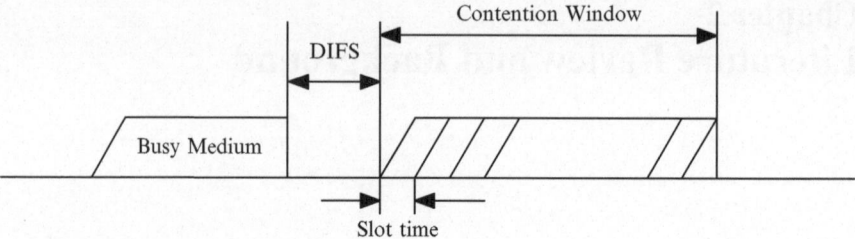

Fig. 2.1 An illustration of IEEE 802.11 DCF

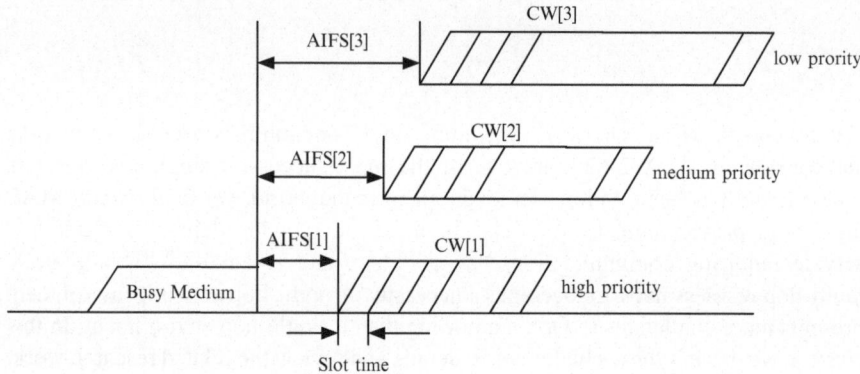

Fig. 2.2 An illustration of IEEE 802.11 EDCA

each successful transmission. On the other hand, with PCF, a contention-free period (CFP) and a contention period (CP) alternate periodically. During CFP, access point (AP) polls stations to grant a transmission opportunity to each station. When polled, a station transmits its frames without collision. The main drawbacks of PCF include uncontrolled transmission time of polled stations and unpredictable CFP start time [66].

Mainly designed for data transmission, DCF does not take into account the delay-sensitive nature of real-time services and does not provide any differentiated services. Various schemes [5, 9, 22, 60] have been proposed to modify IEEE 802.11 DCF to incorporate differentiated services. Summarizing the common feature of those schemes, the IEEE LAN/MAN Standards Committee develops IEEE 802.11e [3] to enhance the legacy IEEE 802.11 MAC with QoS provisioning to real-time applications. As an extension of DCF, the enhanced distributed channel access (EDCA) provides a priority scheme to differentiate different access categories (ACs) by classifying the arbitration interframe space (AIFS), and the initial (CW_{min}) and maximum (CW_{max}) contention window sizes in the backoff procedures, as shown in Fig. 2.2. High priority traffic (e.g., real-time voice) is assigned smaller AIFS than low priority traffic so that it waits for a shorter time before counting down its backoff timer. High priority traffic is also assigned smaller CW_{min} and CW_{max} values than low priority traffic, so it has more chances to choose a smaller backoff timer and counts down to zero earlier, thus gets the channel earlier.

2.1.2 Limitations of IEEE 802.11 in QoS Support

DCF/EDCA is not effective or efficient in supporting the delay-sensitive voice traffic. Both experimental results [31,91] and analytical results [15,37] demonstrate that system capacity for voice traffic is very limited in WLANs due to the large header overhead and the inefficiency of IEEE 802.11 MAC protocol. The time to transmit the payload of a voice packet is only a very small portion of the total time to transmit the packet, due to the large overhead such as the RTP/UDP/IP (Real-time Transport Protocol/User Datagram Protocol/Internet Protocol) headers, MAC header, physical preamble, the IFSs, and the backoff time. Consequently, the capacity to accommodate voice traffic in DCF or EDCA is very limited. For example, IEEE 802.11b can support approximately 11 simultaneous two-way voice calls if a GSM (global system for mobile communications) 6.10 codec is used [91]. In addition, EDCA provides only statistical rather than guaranteed priority access to voice traffic. In other words, the priority for voice traffic is only guaranteed in a long term, but not for every contention. Since each station[1] continues to count down its backoff timer once the channel becomes idle for an IFS, a low priority packet (e.g., a data packet) with a probably large initial backoff timer will eventually count down its backoff timer to a small value, most likely smaller than the backoff timer of a newly backlogged high priority packet (e.g., a voice packet). Then the low priority packet grabs the channel, resulting in the high priority packet waiting for a long time for the next competition [99]. Such statistical priority access is hard to satisfy the delay requirement of each voice packet. Furthermore, when applying EDCA, with an increase of low priority traffic load, the collision probability seen by the high priority traffic increases. High priority traffic can suffer from performance degradation due to low priority traffic offering heavy load [72].

Although IEEE 802.11e can provide a certain degree of service differentiation, only very limited traffic classes (i.e., four classes) are supported. All the data traffic receives the same best-effort service without further class differentiation, which may not meet the different requirements of users.

2.1.3 Related Work Review

In order to improve the capacity of voice traffic over WLANs, various solutions have been proposed [13, 36, 48, 91, 97, 105]. A cyclic shift and station removal polling scheme is proposed in [105] to take advantage of multiplexing voice packets. Without changing the IEEE 802.11 MAC protocol, VoIP capacity is increased by reducing the header overhead of voice packets in [91, 97]. A voice multiplex-multicast (M-M) scheme is proposed in [91], in which the AP multiplexes packets from several VoIP streams into one multicast packet for transmission.

---

[1]In this book, the words "node" and "station" are used interchangeably.

However, an additional delay is expected in composing such a composite packet. In [97], compressed RTP is used to reduce the VoIP header; however, the overhead incurred by IP, MAC and physical layer remains high. On the other hand, some research [13, 36, 48] increases the capacity by introducing new MAC protocols. By reducing the number of collisions or reducing the idle time caused by backoff, these new MAC protocols achieve a better throughput than the IEEE 802.11 MAC protocol, resulting in an increased capacity.

Since the system capacity is very limited in WLANs, call admission control is important and necessary to maintain the QoS of existing calls. As revealed in [31], an additional call that exceeds the system capacity will cause unacceptable quality for all ongoing calls. Previous research on call admission control in WLANs can be classified into two categories. One is analysis based, and the other is measurement based. Analysis based admission control algorithms, including [27, 52] and the reference model provided in IEEE 802.11e, make an admission decision based on the knowledge of the system capacity derived from analysis. The measurement based algorithms [30, 70, 97] make an admission decision based on the measurement or estimate of channel utilization. Based on the measurements of the fraction of time per time unit needed to transmit the flow over the network [30], collision statistics of each flow [70], or the transmission time of each traffic type [97], available/residual budgets are calculated for admission control.

Most of previous work assumes that the voice traffic is constant rate traffic, which is not the case in reality. For more accurate capacity estimate, the on/off model [73] (to be explained in Sect. 2.3) should be applied and the voice traffic multiplexing should be considered. Further, most of the previous work focuses on contention-based access (e.g., EDCA). However, contention-based access can not provide bounded delay performance to delay-sensitive voice traffic. On the contrary, controlled access can provide guaranteed delay performance for voice traffic, making it more suitable for voice traffic delivery. Unfortunately, only very limited work focuses on controlled access. The capacity of PCF is analyzed in [18, 81].

Most of the research in the literature focuses on the QoS support for real-time traffic, and the QoS of data traffic is seldom considered. To the best of our knowledge, there is no work to address the issue of providing further class differentiation to data traffic according to different user requirements in WLANs.

2.2   MAC in Multi-hop Wireless Networks

2.2.1   Problems Due to a Multi-hop Wireless Environment

The IEEE 802.11 MAC protocol has gained great success in WLANs, and also becomes a popular protocol for wireless ad hoc networks. However, several problems (described in the following, which do not exist in a single-hop network) may occur when applying the IEEE 802.11 protocol in a multi-hop network, leading

Fig. 2.3 Problems of
hidden/exposed terminal,
priority reversal, and
unfairness

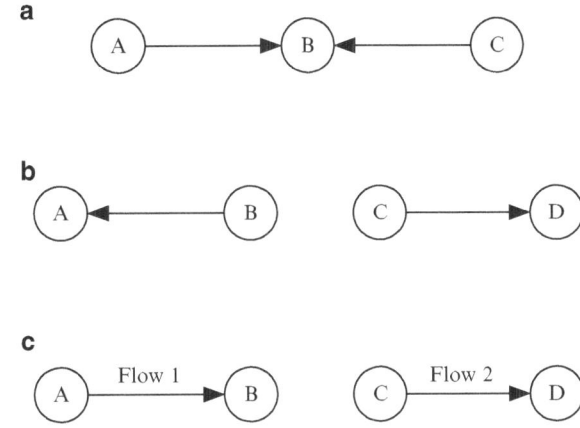

to an unsatisfactory performance. Note that these problems are not just associated
with IEEE 802.11. Many CSMA/CA based MAC schemes (e.g., [5, 9, 22]) that use
a backoff mechanism similar to that of IEEE 802.11 may face the same problems.
Without solving all these problems, QoS provisioning for multimedia applications
in multi-hop wireless networks is difficult to achieve.

Hidden terminal problem – An example of the hidden terminal problem is shown
in Fig. 2.3a, where nodes A and C are out of the carrier sense range[2] of each other,
and both want to send data to the common neighbor B (called two nodes neighbours
if they are within the transmission range[3] of each other). Suppose node A starts to
transmit to node B first. When node C senses the channel and finds the channel idle
(since it is out of the carrier sense range of node A), it will start its transmission and
collide with node A's transmission. Node C is called a hidden terminal of node A.
Hidden terminals cause collisions, leading to wasted resources and a reduced system
throughput.

Exposed terminal problem – An example of the exposed terminal problem is
shown in Fig. 2.3b, where each node can hear its immediate neighbors but no other
nodes. Suppose node B is sending packets to node A and, at the same time, node C
wishes to send packets to node D. Node C senses the channel and finds the channel
is busy, thus defers its own transmission, even though node C's transmission does
not interfere with the reception of node A. Node C is called an exposed terminal of
node B. Exposed terminals cause inefficiency in channel utilization.

Priority reversal problem – An example of the priority reversal problem is shown
in Fig. 2.3c, where flow 1 (from node A to node B) has a higher priority than flow 2

---

[2]Carrier sense range of a node is the range within which other nodes can detect a busy channel
when the node is transmitting.

[3]Transmission range of a node is the range within which all other nodes can correctly receive
packets transmitted from the node in the absence of interference.

(from node C to node D). Flow 1 and flow 2 conflict with each other since node B and node C are neighbours. It is likely that flow 1 may lose its priority when competing with flow 2. The reason is that node C is a hidden terminal of node A and cannot be aware of the transmission of node A. Even though node A may start its transmission earlier than node C, it is possible that node C starts its own transmission before node A finishes the transmission (i.e., their transmission times overlap), resulting in a collision at node B. The reception at node D is successful in this scenario. As a result, the low-priority flow (i.e., flow 2) but not the high-priority flow (i.e., flow 1) delivers its frame successfully.

Furthermore, in a multi-hop environment, it is possible that a high-priority flow with more neighbours experiences a higher contention degree[4] than its low-priority contenders with less neighbours. As a result, the priority access cannot be ensured either [99].

Serious long-term unfairness problem – Although the IEEE 802.11 MAC protocols are characterized by inherent short-term unfairness [51], they do have a good performance of long-term fairness[5] in a fully-connected network. However, when applied to a multi-hop environment, they introduce serious long-term unfairness (some flows may be starved). Consider the same topology shown in Fig. 2.3c, where flow 1 and flow 2 have the same priority. When IEEE 802.11 is deployed, flow 1 is almost starved and flow 2 occupies almost all the channel capacity for the following reason. When nodes A and C both send packets to their destination B and C, respectively. Node B cannot correctly receive node A's packet due to the interference from node C. Without getting the response from node B, node A will double its contention window size and retransmit. Its contention window will eventually reach the maximum value CW_{max} after a number of retransmissions. On the other hand, if node A is transmitting, node C knows exactly the finish time of node A's transmission (by overhearing node B's CTS, to be explained later), thus defers its own transmission until node A finishes its transmission. Hence, node C maintains the minimum contention window size CW_{min}. As a result, node A is unlikely to get the channel (due to the large contention window size) and will get starved.

2.2.2 MAC over Wireless Ad Hoc Networks

In the literature, many MAC schemes have been proposed for wireless ad hoc networks to avoid the above problems (a literature survey is given in [46]). To avoid the collisions caused by hidden terminals, the request-to-send (RTS)/clear-to-send

---

[4]Contention degree of a flow is defined as the numbers of flows with which the flow is competing for the channel.

[5]For long-term fairness, fair service shares are achieved among all the contending nodes in a relative large time scale (e.g., 10 s). On the other hand, short-term fairness should be achieved in a small time scale (e.g.,10 ms).

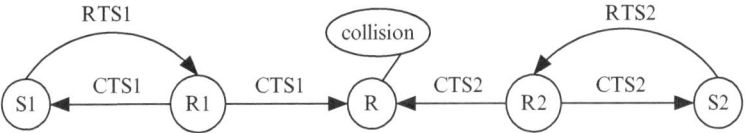

Fig. 2.4 An example of CTS collision

(CTS) approach is widely adopted. Examples include MACA (Multiple Access Collision Avoidance) [47], MACAW (MACA for Wireless LANs) [10], and IEEE 802.11 DCF [1]. A node that has data to send first sends an RTS, which carries the information of the amount of time that the channel will be utilized to complete the following data packet transmission. All nodes hearing the RTS will defer their transmissions. If the intended receiver grants the request, it will return a CTS, which also indicates the time needed to complete the reception of the upcoming data packet. All nodes hearing the CTS also defer their transmissions. See again the example in Fig. 2.3a. Upon receiving the CTS from node B, node C will defer its own transmission until node A finishes its transmission. Hence, the hidden terminal problem is avoided. However, the RTS/CTS dialog is less effective to avoid collisions in a relatively crowded region because RTS and CTS frames themselves are still subject to collisions [92]. Figure 2.4 shows an example of the CTS collision. The CTS collision does not mean that the sender cannot correctly receive CTS from the intended receiver. It refers to the case that some of the receiver's neighbours cannot correctly receive CTS due to collision. As shown in Fig. 2.4, node R is the neighbour of both R1 and R2, while R1 and R2 cannot hear each other. When two senders S1 and S2 send RTSs simultaneously to R1 and R2, respectively, R1 and R2 will respond with CTS at the same time and cause collision at node R. Therefore, node R has no idea about both transmissions. Later when node R wants to send data to other node, it will initiate its RTS, thus corrupt the data reception at R1 and R2. This CTS collision results in throughput degradation of the RTS/CTS scheme in multi-hop networks.

Another popular approach is to protect the receiver's data packet reception by adding a busy-tone channel which is separated from the information channel [33,78, 102]. In [78], all nodes hearing an ongoing transmission will send a busy-tone in a separated narrowband. Upon hearing a busy-tone, all nodes defer their transmission; therefore, all nodes within two hops of the transmitting source node are prohibited from transmission. In [33, 102], a receiver sends a busy-tone in the narrowband busy-tone channel during the period of the data packet reception to indicate whether or not the receiver is receiving a data packet. upon hearing the busy-tone, all the neighbors of the receiver defer their transmissions. The busy-tone solution avoids data packet collisions; however, RTS frame collisions caused by hidden terminals still exist. According to IEEE 802.11b, the RTS frame size is 20 bytes. Considering a physical layer overhead (192 μs), 272 μs is needed to transmit an RTS frame at the required basic rate (i.e., 2 Mbps). With one slot time equal to 20 μs, one RTS transmission time approximately equals to 14 slots. Suppose two hidden terminals, without suffering from collisions previously, randomly pick up their backoff timers from the initial contention window CW_{min} (31 in IEEE 802.11b) and start to count down their backoff timers simultaneously. The probability that these two nodes' RTSs collide with each other is as high as 74.8% (the probability that, when two integers are randomly chosen from 0 to 31, the distance between them is less than 14). If more than two hidden terminals exist, the collision probability will be higher, resulting in a reduced system throughput. On the other hand, for voice transmissions, due to the small voice packet size, usually no RTS/CTS dialog is adopted. For voice packet transmissions, the collision probability can be higher since a voice packet size is normally larger than an RTS frame size. None of the previous work gives a solution to avoid voice packet collisions caused by hidden terminals.

For the exposed terminal problem, the dual busy-tone multiple access (DBTMA) [33] has been proposed via dual busy-tone channels. However, ACK frames are omitted in DBTMA (which is not reasonable for unreliable wireless channels) because, otherwise, collisions may happen when a sender is receiving an ACK frame while another nearby sender is transmitting a data packet.

For the priority issue, most of the previous work focuses on single-hop networks, and only limited work addresses the characteristics of a multi-hop environment. A contention-based MAC scheme is proposed to provide priority scheduling in multi-hop networks using busy-tones [99]. Instead of using contention-based schemes, reservation based schemes are proposed [45, 56]. Reservation based schemes adopt the idea of asynchronous time-division multiple access (TDMA) [61], where the channel time is partitioned into cycles. The length of the cycle is usually chosen as the minimum arrival interval of real-time traffic packets, and each cycle contains two periods: a reservation request period and a data transmission period. In the reservation request period, all the competing nodes contend for the channel to submit their reservation requests for the data transmission period. Once the reservation request is granted, a node can transmit data during the reserved period without collision. Thus, the data transmission period is conflict-free. For real-time traffic, only the first packet of a call needs to contend for the channel. Once its reservation request is granted, all the subsequent packets in the same call will be conflict-free when transmitted in the reserved period. On the other hand, the non-real-time nodes have to contend for the channel for every data packet. However, it needs extra signaling overhead to exchange the reservation information, and the information maybe inconsistent at different nodes due to collision, location-dependent error, and node mobility.

In [28], a distributed fairness algorithm is proposed for multi-hop wireless backhaul networks,[6] taking the unique network characteristics into account, and the location-dependent unfairness is prevented. However, it is tailored for wireless backhaul networks and may not be suitable for a general multi-hop wireless ad hoc networks. In [39], an ideal fairness model is proposed for multi-hop

---

[6]In wireless backhaul networks, there is an Internet entry point, and all the traffic flows are from and to the Internet entry point via single or multiple hops.

ad hoc networks, taking spatial frequency reuse into consideration. This model requires the information of full network topology, imposing extreme complexity for implementation. In self-coordinating localized fair queueing [58], the service tag information should be exchanged among neighboring nodes, leading to a certain level of information exchange overhead. In [25, 26], a MAC scheme using randomly ranked mini-slots is proposed to maintain fairness in ad hoc networks.

Although some efforts have been made to address some of the problems discussed in Sect. 2.2.1, to the best of our knowledge, so far there is no MAC scheme providing a comprehensive and effective solution for all the problems.

2.2.3 MAC over Wireless Mesh Networks

Although wireless mesh networks can be considered as a type of ad hoc networks, two unique characteristics of the wireless mesh backbone result in that it may not be effective or efficient to directly apply existing MAC schemes proposed for ad hoc networks to the wireless mesh networks [20]. First, many existing MAC schemes for ad hoc networks are designed to handle node mobility with power consumption constraints. For the wireless mesh backbone, the wireless routers are usually located at fixed sites with wired power supply. Thus, the node mobility and power consumption should not be the main consideration for the MAC design. Second, contention-based MAC schemes (e.g., IEEE 802.11) are one major stream for wireless ad hoc networks. However, the traffic volume in the wireless mesh backbone may be much higher than that in an ad hoc network due to traffic aggregating at each router. It is well known that, when traffic load is heavy, contention-based MAC schemes suffer from serious collisions due to the severe contention, leading to dramatically decreased throughput and increased delay.

As pointed out in [7], for application to wireless mesh networks, all existing MAC schemes need to be enhanced or re-invented. So far, very limited work has been done to enhance the existing MAC schemes or design a new MAC scheme specifically for wireless mesh networks. To enhance IEEE 802.11, in [16], an end-to-end reservation protocol is proposed to support QoS of real-time traffic. In [103], a new protocol named Wireless Channel-oriented Ad-hoc Multi-hop Broadband (W-CHAMB) is proposed based on time-division multiple access/time division duplex (TDMA/TDD) technology. In [43], with a cross-layer design principle, an interference aware MAC scheme is proposed for a code-division multiple access (CDMA)-based wireless mesh backbone.

2.3 Traffic Class and QoS Requirements

In this book, both delay-sensitive voice traffic and delay-insensitive data traffic are considered for WLANs and wireless ad hoc networks. Originally designed for high-rate data traffic, WLANs have been reported to have a limited capability to support

Fig. 2.5 The two-state voice
model

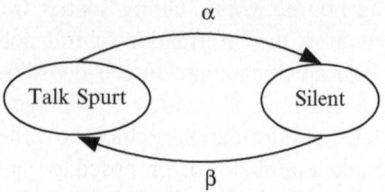

voice traffic. Since video traffic normally requires more bandwidth than voice traffic,
in this study, we do not consider video traffic in WLANs. Wireless ad hoc networks
usually have limited resources (e.g., bandwidth and power), and are not suitable
for video transmissions. Serving as a wireless broadband Internet access network, a
wireless mesh network is expected to support heterogeneous traffic types including
voice, video, and data traffic. Therefore, for wireless mesh networks, all the three
types of traffic are considered. For real-time traffic (voice and video), in addition
to delay, delay jitter (i.e., variation of voice packet delay) should also be carefully
controlled as it may degrade the quality more severely than delay. Traditionally, an
appropriately designed playout buffer is an effective way to deal with delay jitter
and make the real-time traffic understandable. Therefore, delay bound and packet
loss rate guarantees are the main QoS requirements for real-time service under
consideration.

- Voice Traffic: The application of VoIP is considered. VoIP has much lower
  cost than the traditional telephone services, and the emerging digital signal
  processing (DSP) and voice coding/decoding techniques make VoIP more and
  more mature and feasible in voice conversations [12]. VoIP application is delay-
  sensitive but can tolerate a certain level of packet loss. The acceptable end-to-end
  delay of voice traffic, in general, ranges from 150 to 400 ms. Considering the
  delay incurred by coding, decoding and packetizing, the network packet delay
  corresponding to the time taken for a packet to traverse the network, ranges from
  100 to 350 ms [41]. For a good voice quality, the packet loss of compressed
  speech such as GSM 6.10 should be no more than 1% [95]. Generally, voice
  traffic can be represented by an two-state on/off model [73] as shown in
  Fig. 2.5: a voice user is alternately in talk spurt (on state) and in silence
  (off state). The durations of the on and off states are independently and
  exponentially distributed with parameters α and β, respectively. At an on state,
  voice packets are generated periodically, while no voice packet is generated at an
  off state.
- Video Traffic: Video calls generate packets at a time-varying rate according to
  their characteristics and the coding schemes used. In this study, we choose the
  H.264 codec [67], which is the most efficient video compression technology and
  is widely implemented. The H.264 defines a set of profiles with different video
  bit rates for various classes of applications. Video traffic has a wide range of
  bit error rate (BER) requirements, which correspond to several levels of video
  quality. The lower the BER, the better the video quality. The normal tolerable
  delay of video traffic is about 50 ms [65].

- Data Traffic: Data traffic is usually bursty. The number of data packets generated at each burst can vary greatly depending on different data applications. The data applications include email, web browsing, and data transfer applications, etc. The data bursts from a single data source can be modeled by a Poisson process. Data traffic is usually delay-insensitive. However, it can tolerate few transmission errors. We assume that the BER requirement of data traffic can be satisfied by applying automatic repeat request (ARQ) mechanism at the link layer and transmission control protocol (TCP) or other reliable transport protocols at the transport layer. In this research, the QoS metrics to be considered for data traffic are throughput and fairness. The objective is to allocate the network resources to different data flows in a fair and efficient manner.

2.4 Summary

Medium access control plays an important role in supporting QoS over wireless networks. As a WLAN standard, IEEE 802.11 is originally designed for high-rate data traffic and has limited capability to support real-time voice traffic. It has been found out that the system capacity for voice users can be very low in current WLANs due to the large overhead and inefficient channel utilization. Although some research work has been done to address these limitations, most of them focuses on the contention-based MAC, and pay less attention to the controlled MAC. However, the controlled MAC can provide bounded delay performance to the delay-sensitive real-time traffic, while the contention-based MAC has no QoS guarantee. Furthermore, most of the existing MAC schemes for WLANs focuses on the QoS of real-time traffic, and neglect the QoS of data traffic. There is no work to provide further class differentiation to data traffic according to different user requirements. Unlike single-hop WLANs, the multi-hop wireless networks present more challenges to the QoS-aware MAC scheme design because of the existence of the hidden/exposed terminal problem, the location-dependent unfairness, and the priority reversal problem. Although some efforts have been made to address some of these problems, till now there is no scheme providing a comprehensive solution for all these problems.

Chapter 3
Voice Capacity Improvement over Infrastructure WLANs

As discussed in Chap. 2, the current WLANs experience bandwidth inefficiency when supporting voice traffic, leading to a very limited capacity to voice users. In this chapter, we aim at addressing this limitation. Our work is based on IEEE 802.11e since it is the most promising technology for QoS provisioning in WLANs. With minor modifications to IEEE 802.11e, we can increase the system capacity significantly for voice traffic, provide guaranteed QoS to voice users and, at the same time, provide data traffic a certain level of service share.

3.1 Wireless Local Area Network

WLANs are one of the most successful wireless networks, which have been deployed all over the world as a wireless extension to the wired Ethernet. A WLAN is usually a single-hop wireless network and can only cover a small geographic area. For instance, an 802.11b AP can communicate with a mobile user within up to 60 m at 11 Mbps and up to 100 m at 2 Mbps.

A WLAN consists of two modes: infrastructure mode and ad hoc mode [19], as shown in Fig. 3.1. In the infrastructure mode, the AP works as gateway providing an interface to the wireless users in its coverage so that they can access the Internet. Most of the traffic is through the AP. Both centralized MAC (e.g., PCF) and distributed MAC (e.g., DCF) can be applied. In the centralized MAC, AP works as a central controller. In the distributed MAC, all the users as well as the AP contend for the channel. Infrastructure mode is adopted in most of the installed WLANs. In the ad hoc mode, wireless users can spontaneously form a WLAN without the need of AP. All the users can communicate directly with each other in a peer-to-peer manner. The ad hoc mode is suitable for conference meeting and distributed computing, etc. We will study these two modes in Chaps. 3 and 4, respectively.

P. Wang and W. Zhuang, *Distributed Medium Access Control in Wireless Networks*, SpringerBriefs in Computer Science, DOI 10.1007/978-1-4614-6602-4_3, © The Author(s) 2013

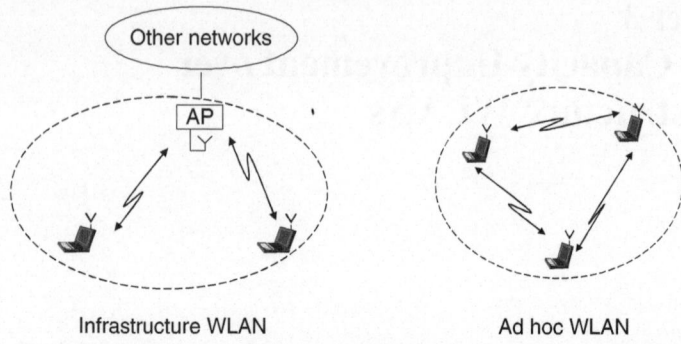

Infrastructure WLAN   Ad hoc WLAN

Fig. 3.1 The infrastructure mode and ad hoc mode of WLANs

3.2 The Service Interval Structure

Time is partitioned into service intervals. We use the same structure of service interval as in IEEE 802.11e. In each service interval, there are two periods: contention-free period (CFP) and contention period (CP), as shown in Fig. 3.2. The CFP is used to accommodate voice stations in the downlink (from the AP to the mobile stations) and uplink (from mobile stations to the AP) by polling. For the uplink transmission, the AP sends a CF-Poll frame which grants each polled station a transmission opportunity (TXOP). No ACK/retransmission is required for voice transmission in order to avoid the retransmission delay. In the CP, the AP and all the stations can contend for the channel. It is mainly used to serve data stations and to transmit the first few packets of each voice station's uplink talk spurts. To guarantee the priority of voice over data in the CP, voice packets are always transmitted ahead of data packets (to be discussed in Sect. 3.3). The length of a service interval is fixed and depends on the delay bound of voice traffic. The length of the CFP and CP depend on the voice and data traffic load and the QoS provisioning technique.

3.3 Mechanisms for Capacity Improvement

The proposed mechanisms for capacity improvement consists of two parts: voice traffic multiplexing and overhead reduction, as elaborated in the following.

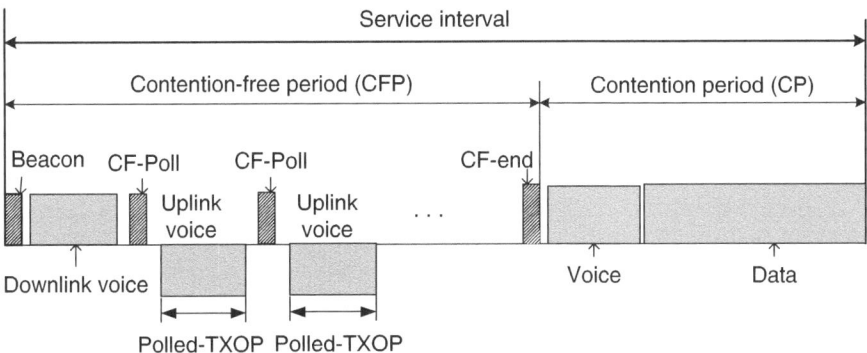

Fig. 3.2 The structure of a service interval

3.3.1 Voice Traffic Multiplexing

3.3.1.1 Dynamic Polling During CFP

In order to achieve a high resource utilization, the network designers should consider the on/off characteristic of voice traffic, so that resources are allocated to stations only when they are in a talk spurt. However, IEEE 802.11e does not provide a method to achieve voice traffic multiplexing. Generally, it is easy for the AP to recognize the ending moment of a talk spurt, but it is difficult to know the exact starting moment of a talk spurt. The AP may still need to poll a voice station even during its silent period in order not to miss the beginning of a talk spurt, which is not efficient considering the polling overhead. Here we propose a more efficient polling mechanism to achieve the voice traffic multiplexing.

Consider the case when a station initiates a voice call. If the call can be admitted, the AP will add the station to its polling list. Since the duration of each service interval (T_{SI}) is fixed and the voice packet inter-arrival time (I_o) is a constant in a talk spurt, each station (in the on state) will be granted a fixed TXOP just enough to accommodate the generated voice packets during a service interval. If a polled station has no packet to send or cannot use up all the time of TXOP, the AP considers the station being in the silent period and deletes it from its polling list, except the newly added (to the polling list) stations. When a previously off station has voice packets to send, the station will contend for the channel during the next CP. Once it gets the channel, it will send out all the voice packets in the buffer (as long as the transmission time does not exceed the TXOP). The AP monitors all the packets transmitted in each CP. For every voice packet, the AP records the sender address (or ID) and adds it to the polling list. If the station is newly added to the list during the last service interval, the AP will retain it in the list, even though it may not use up all the TXOP or has no packet to send in the current CFP, since a few voice packets at the beginning of a talk spurt were sent during the last CP.

Once a voice station is added to the polling list, all the subsequent voice packets in the same talk spurt will be transmitted in the CFP. Hence, the voice station does not need to contend for the channel anymore for the current talk spurt.

3.3.1.2   Guaranteed Access Priority to Voice During CP

Another challenging issue is raised from the uplink voice multiplexing: to meet the strict delay requirement of uplink voice traffic, it should be guaranteed that a voice station can access the channel successfully during the CP when needed.

To provide QoS guarantee for voice traffic regardless of the data traffic load in a WLAN, data stations should not transmit in the CP until no voice station contends for the channel. As discussed before, EDCA cannot meet this requirement. As a result, a guaranteed prioritized access for voice traffic is more appropriate.

A simple way to provide guaranteed prioritized access is to modify EDCA so that the AIFS of data access category (AC) (AIFS[AC_data] = AIFS[AC_voice] + CW_{max} [AC_voice]), the summation of AIFS of voice AC and the maximum contention window size of voice AC. However, it is not efficient in terms of channel utilization. The number of contending voice packets is expected to be small in a CP, and all the data packets have to wait a long time before getting the channel, resulting in a waste of resources.

Inspired by the idea of black-burst contention [75], here we propose an efficient mechanism to provide guaranteed prioritized access to voice, by minor modifications to the IEEE 802.11e EDCA. In our mechanism, the AIFSs for voice traffic and data traffic remain the same as those in EDCA. In addition, the contention behaviors for data stations remain the same as in EDCA. The contention behaviors of voice stations are modified as follows. For a contending voice station, after waiting for the channel to be idle for AIFS[AC_voice], instead of further waiting for the channel to be idle for a duration of backoff time, the voice station will send a busy tone[1] and the length of the busy tone (in the unit of slot time) is equal to its backoff timer. After the completion of its own busy tone, the station monitors the channel for the duration of a slot time. If the channel is still busy (which means that at least one other voice station is sending busy tone), the station will quit the current contention, keep its contention window, choose a backoff timer randomly from its contention window, and wait for the channel to be idle for AIFS[AC_voice] again. Otherwise, the station (which sends the longest busy tone) will send its voice packets. It is possible that two or more voice stations happen to send the same longest busy tone, resulting in a collision. Contention windows of collided stations evolve by the same way as that in EDCA, and each collided voice station chooses a backoff timer randomly from its contention window for the next contention. Since there is no ACK frame sent back to acknowledge the successful voice transmission, it is difficult for the sender

---

[1]A busy-tone signal is a jamming signal which dose not carry any bit information. It is sent in the information channel, without incurring extra hardware cost.

to recognize the collision. To address the problem in our scheme, for the first packet from a voice station received in a CP, the receiver should send back an ACK frame to the sender. The voice sender continues to contend in the CP if no ACK is received.

In a CP, if there exists at least one voice contender, all data stations will sense the busy tone during the AIFS[AC_data] (>AIFS[AC_voice]), and defer their transmissions. When a collision happens between voice stations, the data stations will wait for the channel to be idle for the duration of ACK timeout plus AIFS[AC_data] before they attempt to acquire the channel, which ensures that voice stations will not lose the channel access priority to the data stations even when a collision happens. Furthermore, when all the active (in terms of uplink transmission) voice stations are included in the polling list, the data stations can make full use of the CP resources.

By using the above mechanism, it seems that the waiting time (before getting the channel) of a voice station is larger than that in EDCA, since the voice station with the largest backoff timer instead of the smallest backoff timer (as in EDCA) gets the channel. However, as the number of voice stations contending for the channel simultaneously is very likely to be small, the initial and maximum window sizes for voice AC can be set to small values, so the negative effect of waiting time should be negligible in our mechanism.

3.3.2 Overhead Reduction

To support voice over WLANs, it is important to reduce the overhead and improve the transmission efficiency over the radio link. The large packet header overhead can significantly affect the capacity of the WLAN in supporting voice service. For example, if a GSM 6.10 codec is used, a voice packet payload is 33 bytes while the RTP/UDP/IP headers are 40 bytes. In addition, the physical preamble, MAC header, and control packets all consume bandwidth. As a result, the overall efficiency is less than 3% [91]. Actions need to be taken to alleviate the effect of the overhead.

Recently, various header compression techniques for VoIP have been proposed. The RTP/UDP/IP headers can be compressed to as small as 2 bytes [17, 53]. The compression technique is adopted in our research.

In our proposed scheme, the physical and MAC layer overheads are further reduced by aggregating the buffered voice packets from or to a voice station together and transmitting them by one MAC frame. Take uplink transmissions as an example. The AP polls each voice station periodically after every service interval, which depends on the delay bound of voice traffic. Within each service interval, several voice packets may be generated and buffered by each voice station. In order to increase the efficiency, we combine the payload of these packets together and add a common MAC layer header instead of sending them one by one. It reduces the overall MAC layer header and physical preamble overhead.

3.4 Voice Capacity Analysis

For a WLAN supporting voice/data traffic, we assign a higher priority to voice traffic. The CFP is used to transmit voice traffic; and in the CP, voice traffic has guaranteed priority over data traffic. To provide data traffic a certain level of QoS, it is required that the average service time in each CP for data traffic is at least a pre-specified fraction (ϕ) of the whole service interval. Hence, we need to determine the maximum number of voice sessions,[2] denoted by N^*, that can be supported by the average fraction $(1\text{-}\phi)$ of time (used in both CFP and CP) in each service interval, with the required packet loss rate guaranteed. To derive N^*, we first need to obtain the time required to serve all voice sessions in CFP and CP, respectively.

3.4.1 Time Required to Serve Contending Voice Sessions in a CP

In each CP, if there is any contending voice session, the whole CP time can be partitioned to two portions: the first portion is used by voice sessions to contend and transmit, while the second portion is for data traffic. In the following, the average time needed to serve the contending voice sessions in a CP is derived.

In a CP, consider n voice sessions contending for the channel. For simplicity of presentation, the contention window of each voice session takes values from the set $\{CW_1^c, CW_2^c\}$ where $CW_2^c = 2 \cdot (CW_1^c + 1) - 1$, and at the beginning of each CP, all contending voice sessions are with CW_1^c. Our analysis can be easily extended to cases with three or more choices for the contention window size.[3] Define state (n_1, n_2), where n_1 and n_2 are the numbers of voice sessions with contention windows CW_1^c and CW_2^c, respectively. Hence, in each CP, the initial state is $(n, 0)$. When a voice session contends successfully (i.e., it is the only one with the largest backoff timer), it will leave the contention. If there is a collision (i.e., there are at least two voice stations with the largest backoff timer among all the contending voice stations), the involved voice stations will double their contention window until the maximum contention window size (i.e., CW_2^c) is reached. After each successful transmission or collision, the state will evolve, remaining in the current state, or moving to the next one. The state transition is shown in Fig. 3.3, where the state $(0,0)$ is the absorbing state when all voice sessions are served. There are totally $1 + 2 + \ldots + (n + 1) = \frac{(n+1)(n+2)}{2}$ states. To understand the state transition diagram, we use state $(1, n - 1)$ as an example. Its next state is $(0, n - 1)$ if the voice session

---

[2]In this chapter, a voice session means a two-way voice call.

[3]Note that for the case of k contention window sizes, k-dimensional Markov chain is needed. When k is large, the computation complexity may be high. However, as shown in [42], our scheme does not require a large contention window sizes. A satisfactory performance can be achieved with three contention window sizes (i.e., $CW_{min} = 3$ and $CW_{max} = 15$).

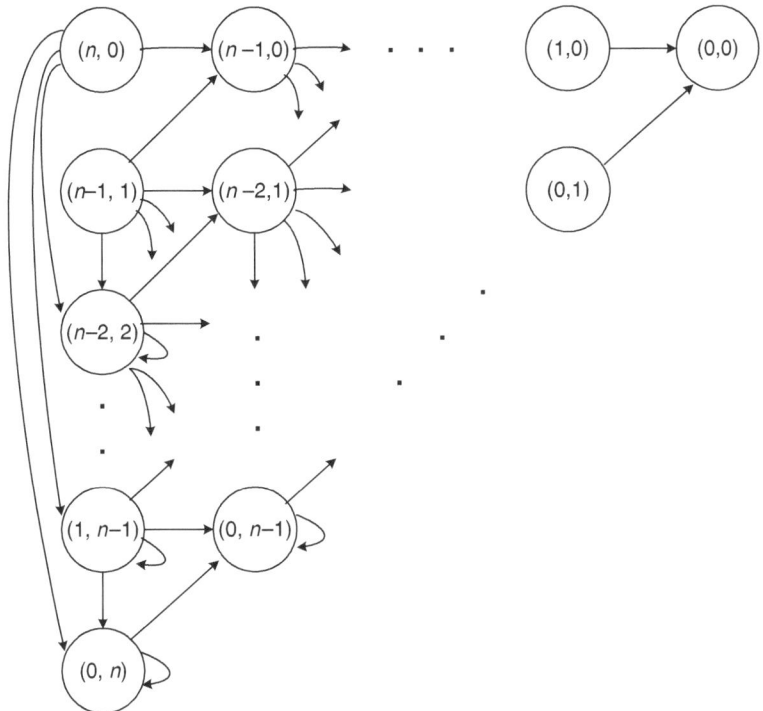

Fig. 3.3 The state transition diagram for (n_1, n_2) in a contention period

with CW_1^c transmits successfully, $(1, n-2)$ if one voice session with CW_2^c transmits successfully, $(0, n)$ if the session with CW_1^c collides with one or more other sessions, or it remains in $(1, n-1)$ if two or more sessions with CW_2^c collide. From the diagram we can also see that the probability of staying in state $(n-1, 1)$ is 0, as no other state enters it.

Let $T(n_1, n_2)$ denote the average time needed for transitions from state (n_1, n_2) to the absorbing state $(0, 0)$. Obviously, we have $T(0, 0) = 0$, and $T(n, 0)$ is the average time to serve all the n contending voice sessions in a CP.

For a state (n_1, n_2), one or more transmissions from voice sessions with either CW_1^c or CW_2^c will lead to its next state. Denote the number of transmissions from voice sessions with CW_1^c and CW_2^c as $l_1(\leq n_1)$ and $l_2(\leq n_2)$, respectively. Denote the next state as $s_{(n_1, n_2; l_1, l_2)}$. Then we have

$$s_{(n_1, n_2; l_1, l_2)} = \begin{cases} (n_1 - l_1, n_2 - l_2) & \text{if } l_1 + l_2 = 1 \\ (n_1 - l_1, n_2 + l_1) & \text{if } l_1 + l_2 > 1 \end{cases} \quad (3.1)$$

where $l_1 + l_2 = 1$ means a successful transmission. When $l_1 + l_2 > 1$, a collision happens, and l_1 involved voice stations originally with CW_1^c will be with CW_2^c after the collision.

Denote the probability of the above transition as $p_{(n_1,n_2;\,l_1,l_2)}$, and the average time of the transition as $t_{(n_1,n_2;\,l_1,l_2)}$. If $l_1 \neq 0$, i.e., the successful transmission or collision happens when the largest backoff timer among all the voice stations takes a value from $[0, CW_1^c]$, we have

$$p_{(n_1,n_2;\,l_1,l_2)} = \sum_{i=0}^{CW_1^c} \binom{n_1}{l_1} \left(\frac{1}{CW_1^c+1}\right)^{l_1} \left(\frac{i}{CW_1^c+1}\right)^{n_1-l_1} \cdot \binom{n_2}{l_2} \left(\frac{1}{CW_2^c+1}\right)^{l_2} \left(\frac{i}{CW_2^c+1}\right)^{n_2-l_2}$$

(3.2)

where the term in the summation represents the probability that l_1 voice stations with CW_1^c and l_2 voice stations with CW_2^c choose a backoff timer value i, and other voice stations choose backoff timer values less than i.

With the condition of the above transition, the conditional probability that the largest backoff timer value in the successful transmission or collision is i can be given by Fall 2009 (Wave 4).

$$\frac{\binom{n_1}{l_1}\left(\frac{1}{CW_1^c+1}\right)^{l_1}\left(\frac{i}{CW_1^c+1}\right)^{n_1-l_1} \cdot \binom{n_2}{l_2}\left(\frac{1}{CW_2^c+1}\right)^{l_2}\left(\frac{i}{CW_2^c+1}\right)^{n_2-l_2}}{p_{(n_1,n_2;\,l_1,l_2)}}$$

and we have

$$t_{(n_1,n_2;\,l_1,l_2)} = \sum_{i=0}^{CW_1^c} \frac{\binom{n_1}{l_1}\left(\frac{1}{CW_1^c+1}\right)^{l_1}\left(\frac{i}{CW_1^c+1}\right)^{n_1-l_1} \cdot \binom{n_2}{l_2}\left(\frac{1}{CW_2^c+1}\right)^{l_2}\left(\frac{i}{CW_2^c+1}\right)^{n_2-l_2}}{p_{(n_1,n_2;\,l_1,l_2)}} \cdot i \cdot \tau + \tau + T_x$$

(3.3)

where τ is the slot duration. On the right side of (3.3), the first term (i.e., the summation) represents the time used by the busy tone, the second term (i.e., τ) is the duration for busy-tone detection after a node finishes its own busy tone, and the third term (i.e., T_x) is the collision or successful transmission time, including the AIFS[AC_voice], the packet transmission time, short interframe space (SIFS), and ACK transmission time for a successful transmission (when $l_1 + l_2 = 1$), or ACK timeout for a collision (when $l_1 + l_2 > 1$).

If $l_1 = 0$, the transmission or collision can happen when the largest backoff timer among all the contending voice stations takes a value from $[0, CW_2^c]$. We have

$$p_{(n_1,n_2;\,0,l_2)} = \sum_{i=0}^{CW_1^c} \left(\frac{i}{CW_1^c+1}\right)^{n_1} \cdot \binom{n_2}{l_2}\left(\frac{1}{CW_2^c+1}\right)^{l_2}\left(\frac{i}{CW_2^c+1}\right)^{n_2-l_2}$$

$$+ \sum_{i=CW_1^c+1}^{CW_2^c} \binom{n_2}{l_2}\left(\frac{1}{CW_2^c+1}\right)^{l_2}\left(\frac{i}{CW_2^c+1}\right)^{n_2-l_2}$$

$$t_{(n_1,n_2;\,0,l_2)} = \left[\sum_{i=0}^{CW_1^c} \frac{\left(\frac{i}{CW_1^c+1}\right)^{n_1} \cdot \binom{n_2}{l_2}\left(\frac{1}{CW_2^c+1}\right)^{l_2}\left(\frac{i}{CW_2^c+1}\right)^{n_2-l_2}}{p_{(n_1,n_2;\,0,l_2)}} \cdot i \cdot \tau \right.$$

$$\left. + \sum_{i=CW_1^c+1}^{CW_2^c} \frac{\binom{n_2}{l_2}\left(\frac{1}{CW_2^c+1}\right)^{l_2}\left(\frac{i}{CW_2^c+1}\right)^{n_2-l_2}}{p_{(n_1,n_2;\,0,l_2)}} \cdot i \cdot \tau \right] + \tau + T_x. \quad (3.4)$$

Hence, consider all possible transitions from state (n_1, n_2) where $n_1 + n_2 > 0$, we have

$$T(n_1, n_2) = \sum_{0 \le l_1 \le n_1,\ 0 \le l_2 \le n_2,\ l_1 + l_2 > 0} p_{(n_1, n_2;\ l_1, l_2)} \left[T\left(s_{(n_1, n_2;\ l_1, l_2)}\right) + t_{(n_1, n_2;\ l_1, l_2)} \right] \quad (3.5)$$

From (3.5) and $T(0,0) = 0$, we can compute the values of $T(n_1, n_2)$.

In addition, to implement this analytical work in a practical system, a lookup table can be generated in advance to reduce the computation complexity in system configuration.

3.4.2 Time Required to Serve Voice Sessions in a CFP

Consider N voice sessions to be served in the WLAN. Each voice session has the independent on and off periods exponentially distributed with mean values $1/\alpha$ and $1/\beta$, respectively. At a time instant, a voice station is at on state with probability $\frac{\beta}{\alpha + \beta}$, and at off state with probability $\frac{\alpha}{\alpha + \beta}$. When a voice station is at on state, the probability that a transition to off state happens after duration t is given by $\exp(-\alpha \cdot t)$. When a voice station is at off state, the probability that a transition to on state happens after duration t is given by $\exp(-\beta \cdot t)$. The maximum number of downlink (or uplink) voice packets generated in a service interval from a voice session is $V_m = T_{SI}/I_o$. For each voice session, let $P_d(i)$ and $P_u(i)$ $(0 \le i \le V_m)$ denote the probability of generating i downlink and uplink voice packets, respectively, for transmission in a CFP. We have

$$P_d(i) = \begin{cases} \frac{\alpha}{\alpha + \beta} [\exp(-\beta \cdot (T_{SI} - i \cdot I_o)) - \exp(-\beta \cdot (T_{SI} - (i - 1) \cdot I_o))] \\ \quad + \frac{\beta}{\alpha + \beta} [\exp(-\alpha(i - 1)I_o) - \exp(-\alpha \cdot i \cdot I_o)] & 1 \le i \le V_m - 1 \\ \frac{\alpha}{\alpha + \beta} [1 - \exp(-\beta \cdot I_o)] + \frac{\beta}{\alpha + \beta} \exp(-\alpha(T_{SI} - I_o)) & i = V_m \\ 1 - \sum_{i=1}^{V_m} P_d(i) & i = 0 \end{cases}$$

$$(3.6)$$

and

$$P_u(i) = \begin{cases} \frac{\beta}{\alpha + \beta} [\exp(-\alpha(i - 1)I_o) - \exp(-\alpha \cdot i \cdot I_o)] & 1 \le i \le V_m - 1 \\ \frac{\beta}{\alpha + \beta} \exp(-\alpha(T_{SI} - I_o)) & i = V_m \\ 1 - \sum_{j=1}^{V_m} P_u(j) & i = 0. \end{cases} \quad (3.7)$$

We use the example of $1 \le i \le V_m - 1$ to explain the above equations. For the downlink, a voice session will generate i packets when it is originally (at the beginning of the service interval) off with probability $\frac{\alpha}{\alpha + \beta}$ and transits to the on state within $[T_{SI} - i \cdot I_o, T_{SI} - (i - 1) \cdot I_o)$ with probability $[\exp(-\beta \cdot (T_{SI} - i \cdot I_o)) - \exp(-\beta \cdot (T_{SI} - (i - 1) \cdot I_o))]$, or when it is originally on with probability $\frac{\beta}{\alpha + \beta}$ and transits to the off state within $((i-1)I_o, i \cdot I_o]$ with probability $[\exp(-\alpha(i-1)I_o) - \exp(-\alpha \cdot i \cdot I_o)]$. The uplink case is different from the downlink case, as the first several packets in each talk spurt (when an off to on transition happens within T_{SI}) in the uplink are transmitted in the CP.

Next we estimate the number (X) of voice packets that can be supported in each CFP. We call the one-way (i.e., uplink or downlink) packets of a voice session ready for transmission in the CFP *a burst* (which will be transmitted by a single MAC frame). For a burst in a CFP, the probability that it is an uplink transmission with size $i(1 \le i \le V_m)$ is

$$\frac{P_u(i)}{\sum_{i=1}^{V_m} P_d(i) + \sum_{i=1}^{V_m} P_u(i)}$$

and the probability that it is a downlink transmission with size $i(1 \le i \le V_m)$ is

$$\frac{P_d(i)}{\sum_{i=1}^{V_m} P_d(i) + \sum_{i=1}^{V_m} P_u(i)}.$$

Then the probability that it is in uplink transmission (thus requiring a CF-Poll) is

$$\frac{\sum_{i=1}^{V_m} P_u(i)}{\sum_{i=1}^{V_m} P_d(i) + \sum_{i=1}^{V_m} P_u(i)}.$$

The average burst size is given by

$$B = \sum_{i=1}^{V_m} \frac{P_u(i)}{\sum_{j=1}^{V_m} P_d(j) + \sum_{j=1}^{V_m} P_u(j)} \cdot i + \sum_{i=1}^{V_m} \frac{P_d(i)}{\sum_{j=1}^{V_m} P_d(j) + \sum_{j=1}^{V_m} P_u(j)} \cdot i. \qquad (3.8)$$

The average number of bursts is X/B.

We have

$$T_{CFP} = (X/B) \cdot (T^o + \frac{L_v \cdot B}{R}) + (X/B) \frac{\sum_{i=1}^{V_m} P_u(i)}{\sum_{i=1}^{V_m} P_d(i) + \sum_{i=1}^{V_m} P_u(i)} \cdot T_{poll} \qquad (3.9)$$

where T_{CFP} is the duration of CFP, T^o is the overhead due to IFS, physical preamble, and MAC overhead, L_v is the payload size of a voice packet, R is the transmission rate of voice payload, and T_{poll} is the polling overhead. Then

$$X = B \cdot \frac{T_{CFP}}{T^o + \frac{L_v \cdot B}{R} + \frac{\sum_{i=1}^{V_m} P_u(i)}{\sum_{i=1}^{V_m} P_d(i) + \sum_{i=1}^{V_m} P_u(i)} \cdot T_{poll}}. \qquad (3.10)$$

Let X_i denote the total number of up- and downlink voice packets from the ith voice session ready for transmission in the CFP of a service interval, and $Y = \sum_{i=1}^{N} X_i$, where N is the total number of voice sessions. The expectation $E[X_i]$

and variance $\text{Var}[X_i]$ of X_i can be determined based on the on/off model. If the packet loss rate in the CFP is required to be bounded by P_L, the following inequality should hold:

$$\frac{\sum_{y>X}(y-X)P_Y(y)}{E[Y]} \leq P_L \tag{3.11}$$

where $P_Y(y)$ is the probability mass function of Y. According to the central limit theorem, the random variable Y can be approximated as a Gaussian random variable with mean $N{\cdot}E[X_i]$ and variance $N{\cdot}\text{Var}[X_i]$ when N is large. The maximum N satisfying the above inequality (3.11) is the maximum voice session number supported by CFP.

3.4.3   Voice Capacity

In Sect. 3.4.1, we derive the average time required in a CP to serve a fixed number, n, of voice sessions contending in the CP. However, with N voice sessions in service, the number of contending voice sessions in a CP varies (due to the voice on/off nature), so does the required service time in the CP. The average service time for contending voice sessions in a CP is given by

$$\overline{T_{CP}^v}(N) = \sum_{n=1}^{N} \binom{N}{n}(P_c)^n \cdot (1-P_c)^{N-n} \cdot T(n,0) \tag{3.12}$$

where $T(n,0)$ is the average time to serve n contending voice sessions in a CP (as defined in Sect. 3.4.1), and P_c is the probability that a voice station contends for the channel in a CP, given by

$$P_c = \frac{\alpha}{\alpha+\beta}[1-\exp(-\beta \cdot T_{SI})]. \tag{3.13}$$

The duration of the CP in a service interval, $T_{CP} = T_{SI} - T_{CFP}$, is larger than $\overline{T_{CP}^v}(N)$, as the difference of them is the average service time for data traffic in a CP. The performance of contending voice sessions in a CP can be evaluated by the outage probability that T_{CP} is not sufficient to serve all the contending voice sessions

$$\xi(N) = \sum_{n=1}^{N} \binom{N}{n}(P_c)^n \cdot (1-P_c)^{N-n} \cdot I\{T(n,0) > T_{CP}\} \tag{3.14}$$

where $I\{\cdot\}$ is an indicator function.

From the analysis in Sect. 3.4.2, if N voice sessions are admitted, we can determine the minimum value of T_{CFP}, denoted by $T_{CFP}^m(N)$, in order to guarantee the voice packet loss rate bound, the voice capacity region N^* should satisfy

$$T_{CFP}^m(N) + \overline{T_{CP}^v}(N) \leq (1-\phi)T_{SI} \tag{3.15}$$

and the service interval should be configured with a CFP with duration $T_{CFP}^m(N^*)$ and a CP with duration $T_{SI} - T_{CFP}^m(N^*)$.

3.5 Numerical Results and Discussion

To validate the analysis and evaluate the performance of our proposed scheme, computer simulations are carried out using Matlab. The simulation for each run consists of 1,000 service intervals. We choose the GSM 6.10 codec as the voice source as an example. The voice payload size is 33 bytes and the packet inter-arrival period is 20 ms. Compressed RTP/UDP/IP headers with size 4 bytes are used in all the simulations. Other simulation parameter values are listed in Table 3.1. We first vary the contending voice session number (i.e., n) in a CP (where each voice session has one voice packet to send), and analyze and/or simulate the time to serve all the voice packets (i.e., the time to serve the contending voice sessions in a CP). Then we evaluate how the packet loss rate in a CFP changes with the number of voice sessions N. In the evaluation, the first several packets of an uplink talk spurt are not transmitted in the CFP (but are transmitted in the CP by contention). Finally we evaluate the capacity of the whole system, and compare it with that of IEEE 802.11e. We obtain the portion of time required to serve different number of admitted voice sessions, and obtain the system capacity.

3.5.1 Time to Serve Contending Voice Calls in a CP

For uplink voice transmission in our scheme, the first several packets in each talk spurt are transmitted in the CP. With the system parameters, the probability of a voice session contending in a CP is around 9% according to the analysis. Hence, if the total voice session number is 200, there are on average 18 voice sessions contending in each CP. Figures 3.4 and 3.5 show the average time required to serve

Table 3.1 Simulation parameters used for infrastructure WLANs

Parameter	Value	Parameter	Value
Slot time τ	20 μs	R	11 Mbps
T_{SI}	100 ms	Basic rate	2 Mbps
SIFS	10 μs	$1/\alpha$	352 ms
AIFS[AC_voice]	40 μs	$1/\beta$	650 ms
AIFS[AC_data]	60 μs	I_o	20 ms
Physical preamble	192 μs	L_v	33 bytes
MAC header	34 bytes	Data packet payload	1,000 bytes
ACK	14 bytes	P_L	1%
CF-poll	36 bytes		

Fig. 3.4 Average time required to serve contending voice sessions in a CP in our scheme

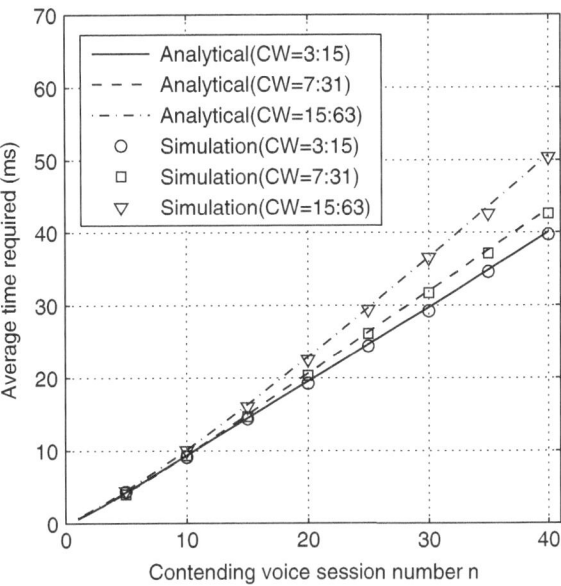

Fig. 3.5 Average voice collision number (via simulations) in a CP in our scheme

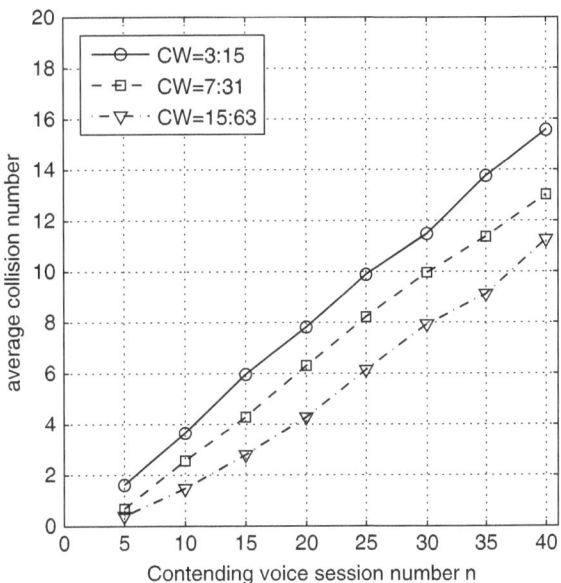

contending voice sessions (i.e., $T(n,0)$) and average voice collision number in a CP versus the contending voice session number n with different settings for initial and maximum contention windows (CW_{min} and CW_{max}), respectively. It is clear that our analysis matches well with the simulations. Contention window settings are critical for contention-based channel access. In our scheme, when the voice sessions have

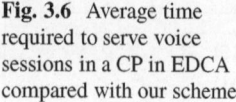

Fig. 3.6 Average time required to serve voice sessions in a CP in EDCA compared with our scheme

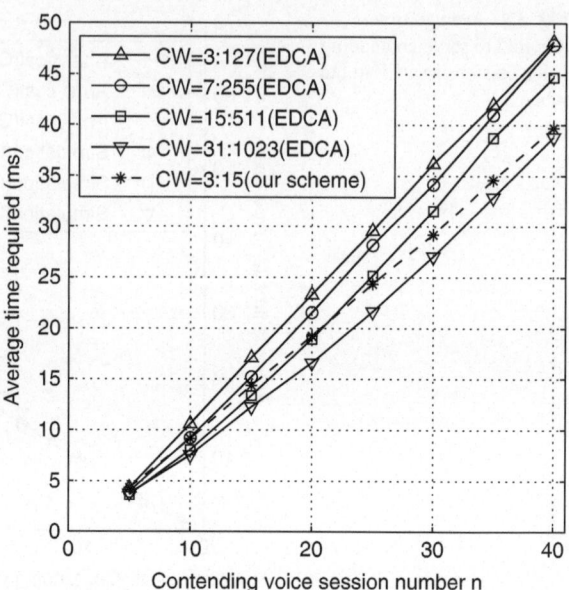

smaller CW_{min} and CW_{max}, the time to transmit busy tone is smaller, at the cost of more collisions. Via the analysis and simulations, we find out that $CW_{min} = 3$ and $CW_{max} = 15$ can lead to the minimal average time to serve all the voice sessions.

Via simulations, we also compare this best case with the cases if EDCA of IEEE 802.11e is applied in the CP, and demonstrate the results in Fig. 3.6. It can be seen that there is no much difference between our scheme and EDCA. Only EDCA with $CW_{min} = 31$ and $CW_{max} = 1,023$ has a non-trivial gain over our proposed scheme. However, to obtain priority in EDCA, voice AC is very likely to have a smaller CW_{min} (<31) and CW_{max} (<1,023). Although a voice station with the largest backoff timer instead of the smallest one (as in EDCA) gets the channel in our scheme, voice performance is not degraded much. The reason is that our proposed scheme can use very small CW_{min} and CW_{max}, but EDCA cannot. If our proposed scheme and EDCA use the same CW_{min} and CW_{max}, the backoff waiting time in our scheme is larger. However, our scheme has a smaller collision probability. If multiple nodes choose the same backoff timer, a collision will occur in EDCA, but a collision will happen in our scheme only when the multiple nodes are with the largest backoff timer (among all the nodes). Figures 3.7 and 3.8 show the average collision number and required time to serve contending voice sessions, respectively, in EDCA and our scheme with $CW_{min} = 3$ and $CW_{max} = 15$ via simulations. It can be seen that, as the contending voice session number increases, the collision number increases rapidly in EDCA, but relatively slowly in our scheme. Hence, the time required to serve contending voice sessions in our proposed scheme is much smaller than that in EDCA when the contending session number is large in the example.

Fig. 3.7 Average collision number in a CP when our scheme or EDCA is applied with $CW_{min} = 3$ and $CW_{max} = 15$

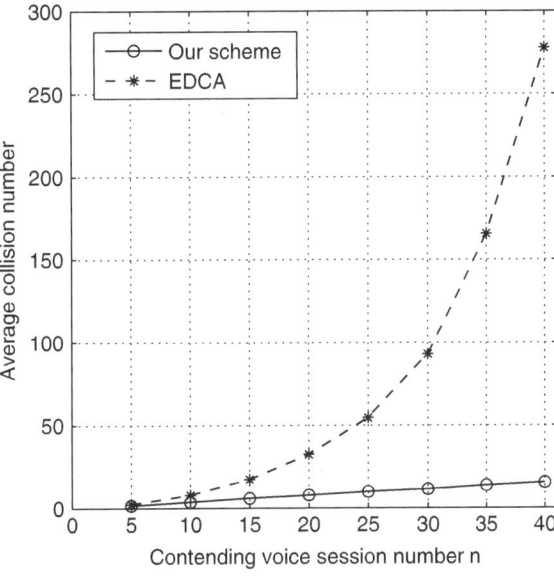

Fig. 3.8 Average time required to serve contending voice sessions in a CP when our scheme or EDCA is applied with $CW_{min} = 3$ and $CW_{max} = 15$

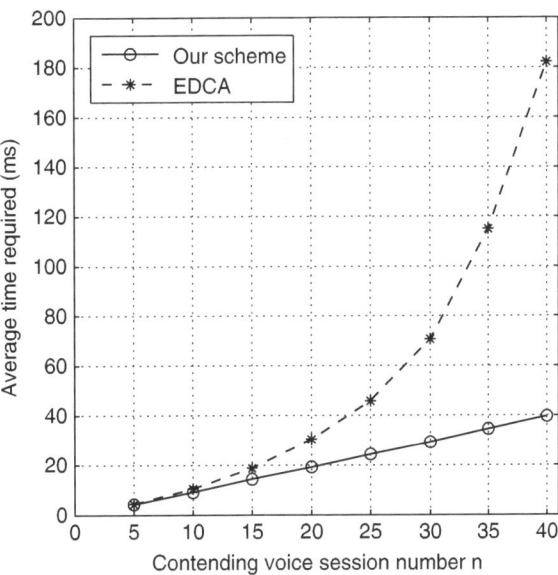

Figure 3.9 shows the effect of data traffic on the average time required in a CP so that all contending voice sessions can be served in EDCA. Long-lived data sessions use the initial and maximum contention window pair (31, 1,023), while voice sessions choose initial contention window size 15 and maximum contention window size 63, 127, or 255. We can see that when the number of data sessions

Fig. 3.9 Average time required so that all contending voice sessions can be served in a CP when EDCA is used to support 10 voice sessions and variable number of long-lived data sessions with CW[data] = 31:1,023

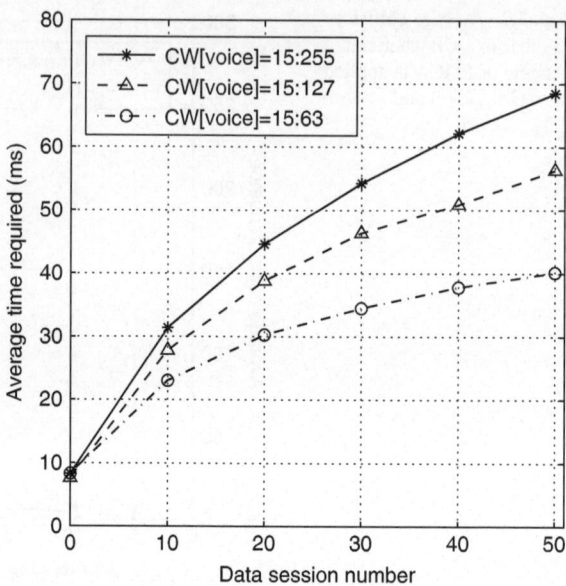

increases, the average time required increases accordingly, and the negative effect is more significant if voice sessions choose a larger initial and maximum contention window pair ((15, 255) in the example).

3.5.2 Packet Loss Rate in CFP

Figure 3.10 shows the analytical results of the packet loss rate versus voice session number N with T_{CFP} equal to 60 and 70% of T_{SI} in our scheme. Simulations are also carried out for selected values of N. It can be seen that our analysis matches well with the simulations. From Fig. 3.10, when the voice session number is equal to or less than 141 when $T_{CFP} = 60\%T_{SI}$ or 167 when $T_{CFP} = 70\%T_{SI}$, the packet loss rate in the CFP is bounded by 1%.

3.5.3 Capacity Region of Voice

To determine the capacity region of voice in our scheme, we vary the number N of voice sessions in the system, and calculate the average time in the CP $\overline{T_{CP}^v}(N)$ and the duration of the CFP $T_{CFP}^m(N)$ in order to guarantee that voice packet loss rate in the CFP is bounded by 1%. We further obtain the total average time in a service interval needed to serve the N voice sessions with QoS guarantee. The analytical results are shown in Fig. 3.11, which also gives the outage probability that the CP with duration

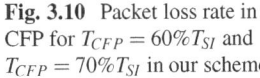

Fig. 3.10 Packet loss rate in CFP for $T_{CFP} = 60\%T_{SI}$ and $T_{CFP} = 70\%T_{SI}$ in our scheme

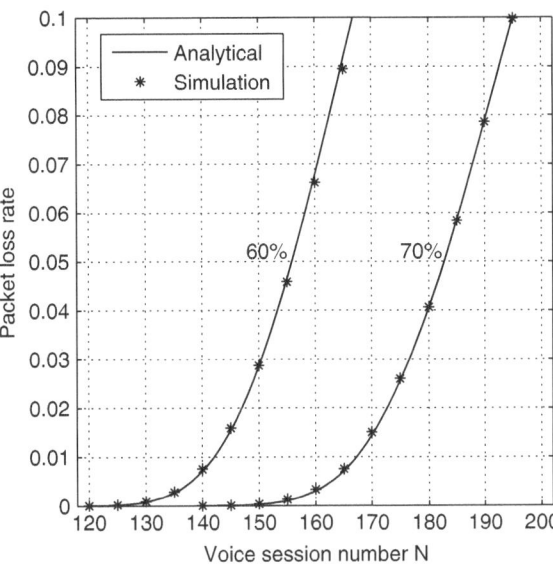

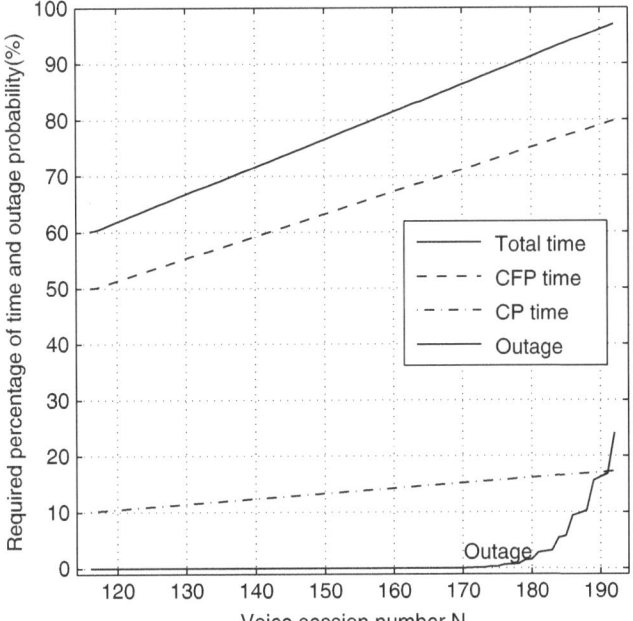

Fig. 3.11 The percentage of time (in CP, CFP, and totally) in a service interval needed to serve the voice sessions with QoS guarantee and the outage probability that not all the contending voice sessions can be served in a CP

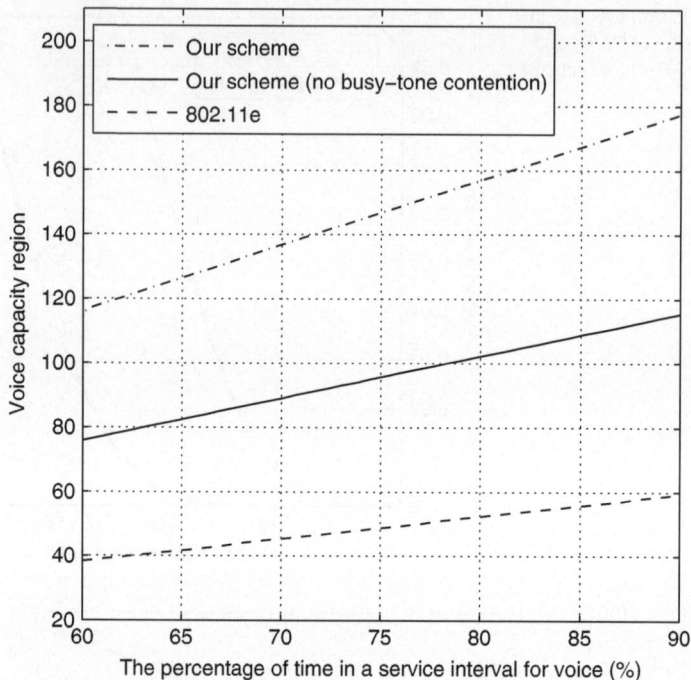

Fig. 3.12 The analyzed voice traffic capacity region of our proposed scheme, our scheme without busy-tone contention mechanism, and IEEE 802.11e

$[T_{SI} - T_{CFP}^m(N)]$ cannot serve all contending voice sessions. It is shown that, when data traffic requires average 30% service time (thus 70% time for voice) in a service interval, we should configure $T_{CFP} \approx 57\%T_{SI}$ and $T_{CP} \approx 43\%T_{SI}$ with a maximum admitted voice session number of 136. The outage probability is negligible ($<1\%$) if the total average time for voice is less than 90%.

We further obtain the voice capacity region (i.e., the maximum number of voice sessions that can be admitted) when the percentage of time (in both CFP and CP) used by voice in each service interval varies from 60 to 90%, and get the analytical results as shown in Fig. 3.12. The analyzed voice capacity region of the IEEE 802.11e polling scheme with the same percentage of time for voice is also included in Fig. 3.12. For a comparison, Fig. 3.12 also shows the voice capacity region when our scheme is applied without the busy-tone contention mechanism (i.e., with only the overhead reduction mechanism). For uplink transmissions, all voice stations are polled, and if a polled voice station has no packets to transmit, it will respond with a NULL frame. From Fig. 3.12, it can be seen that our proposed overhead reduction and busy-tone contention mechanisms both can significantly improve system capacity as compared with IEEE 802.11e.

3.6   Summary

To support real-time voice traffic as well as data traffic over WLANs, the controlled channel access is preferred to the contention-based access for voice traffic. In this chapter, we propose solutions to enhance QoS provisioning capability of IEEE 802.11e to guarantee the delay requirement of voice and, at the same time, ensure data traffic a certain level of service share. Voice statistical multiplexing is exploited effectively, and the system overhead is reduced significantly. Our solutions are shown to significantly improve the voice service capacity of IEEE 802.11e WLANs. An analytical model is also presented to derive the voice service capacity. This research should provide helpful insights to the development and deployment of VoIP technologies over WLANs.

Chapter 4
Service Differentiation over Ad Hoc WLANs

In Chap. 3, an infrastructure-based WLAN is considered with a central controller. In this chapter, we consider an ad hoc WLAN where a distributed access control is needed due to the lack of central controller. Most of the existing distributed access control schemes for ad hoc WLANs are contention-based schemes, which only provide limited service differentiation between real-time traffic and data traffic, and are difficult to provide further service differentiation to data traffic based on some specific requirements of customers or network service providers. In this chapter, we introduce a novel token-based distributed MAC scheme which not only achieves service differentiation between real-time traffic and data traffic, but also achieves proportional class differentiation to data traffic.

4.1 Proportional Class Differentiation Model

Proportional class differentiation model [23] assures that the performance (e.g., throughput) of a class is proportional to that of another class according to a ratio preset by the network service provider. Specifically, consider a WLAN supporting k classes and class s ($s = 1, 2, \ldots, k$) having N_s source nodes. The class differentiation ratio of any two classes, i and j, is denoted by C_i/C_j, i.e., when a node in class i gets C_i fraction of the channel time, a node in class j should get C_j fraction of the channel time. Normalized with the capacity share of class 1, the class differentiation parameter of class s is defined as r_s ($= C_s/C_1$). Although the actual service performance of each class may vary with the traffic load, the performance ratios among classes remain constant. Such a feature provides great flexibility and facility to network service providers for service management. For example, when customers are charged differently for different services, it is desired that the received services (or resources) are proportional to what they are charged.

P. Wang and W. Zhuang, *Distributed Medium Access Control in Wireless Networks*,
SpringerBriefs in Computer Science, DOI 10.1007/978-1-4614-6602-4_4,
© The Author(s) 2013

4.2   The Distributed Token-Based MAC Scheme

There are two tokens in the system. One is circulated among voice nodes (referred to as voice token) and the other is circulated among data nodes (referred to as data token). When a node holds the token, it will transmit its packet(s) when the channel is available. For a voice node, after obtaining the voice token, it transmits all its backlogged packets. For a data node, after obtaining the data token, it is assigned a maximum channel occupancy time (which is the same for all the data nodes and is preset as a system parameter), during which the data node can transmit one or multiple packets depending on its packet size and transmission rate. The proposed scheme works in a distributed manner. There is no central controller passing the tokens to others. The current token holder decides which the next token holder is. When a backlogged voice/data node holds the token, it piggybacks the token in its voice/data packet transmission and passes it to the next node. Note that the destination of the voice/data packet and the next token holder may be different. When a data token holder has no packet to transmit or a voice token holder changes from the on state to the off state, the node passes the token directly to the next holder.

4.2.1   Access Priority and Dynamic Token Passing for Voice Traffic

Voice traffic is given a higher access priority than data traffic. Before a voice (or data) token holder transmits its packet(s), it must first wait for the channel to be idle for T_2 (or T_1, where $T_1 > T_2$). If the channel remains idle during T_2 (or T_1), the voice (or data) token holder transmits; otherwise, it waits for the channel to be idle for T_2 (or T_1) again. The shorter T_2 ensures a higher access priority to a voice token holder. When the voice token holder starts to transmit, the data token holder will sense a busy channel during T_1, and defer its transmission. Note that when a voice node receives the token, it is possible that its packet buffer is empty. This case may happen when the voice traffic load is low. Since the voice packets arrive periodically at a constant rate, it is possible that a node receives the token before the next expected packet arrives. In this case, the voice node holds the token till its next packet arrives, then when the channel is idle for T_2, it transmits the packet and passes the token to another node. Before the packet arrival of the voice token holder, the data token holder will sense an idle channel during T_1, and start its transmission. Figure 4.1 illustrates an example of voice/data transmissions.

Considering the on/off characteristic of voice traffic, a voice node has no packet to transmit during an off period. In order to utilize the channel efficiently, it is desired that the voice token is not passed to the nodes which are at the off state. To do so, we let each voice node be aware of its transition point from the on to the off state. Since the voice traffic has a constant arrival rate during an on period,

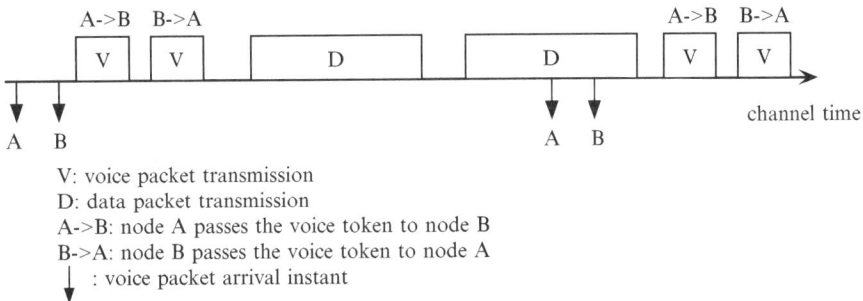

Fig. 4.1   An example of voice and data packet transmissions in the system

the voice node is able to determine when the next expected packet arrives during the period. If the expected packet does not arrive, the node is considered to be at an off state. When a voice node (say node A) holds the token and its state changes from on to off, the node sends a message to announce that it is in the off state and, at the same time, passes the token to the next one (say node B). Upon hearing this message, the previous voice token holder replaces node A with node B as the node to which it will pass the token, so that node A will not receive the token any more during the off period. On the other hand, when a node (say node C) switches from the off state to the on state, it should be able to receive the token to transmit its packets. After waiting for the channel to be idle for T_3 ($< T_2$), node C transmits its packet immediately. In this case, the voice and data token holders will sense a busy channel during T_2 and T_1, respectively, and defer their transmissions. The previous voice token holder monitors the channel after finishing its transmission. Upon hearing node C's transmission, it puts node C as the node to which it will pass the token next time. When node C finishes its transmission, the current voice token holder (say node D) will transmit its packet(s). Upon hearing node D's transmission, node C puts node D as the node to which it will pass the token.

Note that it is possible that a collision may occur when two or more voice nodes (which change from the off state to the on state) transmit simultaneously after the channel being idle for T_3. However, it is shown from both analytical and simulation results that the collision probability is very small and can be neglected. To deal with the rare collisions, p-persistent CSMA [50] can be used. In our scheme, the voice token passing follows a deterministic sequence, which is related to the packet arrival times of voice nodes. Thus the access delay variance is small.

4.2.2   Proportional Class Differentiation Among Data Traffic

The portion of channel time unused by voice nodes is shared by all the data nodes. The data token passing process can be modeled by a stationary Markov chain. Each data node in the WLAN is represented by a state in the Markov chain. The transition probability P_{ij}^t is the probability that node i passes the data token to node j, and the steady-state probability $\pi_t(i)$ of the Markov chain represents the frequency that node i holds the data token compared with others (i.e., on average, if node i holds the data token $\pi_t(i)$ times, then node j will hold the data token $\pi_t(j)$ times). If the following equation is hold for all the data nodes in the network

$$\pi_t(i) = \frac{r_s}{\sum_{s=1}^{k} N_s \cdot r_s}, \quad \text{if node } i \in \text{class } s, s = 1, 2, \ldots, k, \tag{4.1}$$

the channel occupancy ratios achieved by the data nodes in different classes are exactly the same as the required class differentiation ratios ($r_s, s = 1, 2, \ldots, k$), and the channel occupancy time of the data nodes within the same class will be the same.

Given the steady-state distribution, the goal is to find proper values of the transition probability P_{ij}^t of the Markov chain so that the chain's steady-state distribution is exactly the one given in (4.1). According to the Metropolis-Hasting algorithm [34, 62], when we assign a transition probability P_{ij}^t as follows

$$P_{ij}^t = \begin{cases} \frac{1}{N_d - 1} \cdot \min\{1, \frac{\pi_t(j)}{\pi_t(i)}\}, & \text{if } i \neq j \\ 1 - \sum_{k=1, k \neq i}^{N_d} \frac{1}{N_d - 1} \cdot \min\{1, \frac{\pi_t(k)}{\pi_t(i)}\}, & \text{if } i = j \end{cases} \tag{4.2}$$

where N_d is the total number of data source nodes, the corresponding steady-state probability of state i is exactly equal to $\pi_t(i)$ given in (4.1).

It should be pointed out that each data node may have a different transmission rate, depending on the location of the source-destination pair, etc. In this case, even within the same class, a data node with a higher transmission rate will get a higher throughput than those with lower transmission rates. Note that, although the actual throughput may vary, the ratios of the channel time occupied by the data nodes in different classes remain the same as the required ratios. As a result, the system still achieves proportional class differentiation because the resources (i.e., channel time) are proportionally allocated to different classes. With the same amount of assigned resources, a data node may benefit more in term of throughput by increasing its own transmission rate.

Having the network information (e.g., the number of data nodes and their classes), each data node (say node i) first calculates the steady-state distribution $\pi_t(j)$ for any data node j according to (4.1), and then calculates the transition probability P_{ij}^t individually according to (4.2). When node i holds the data token, it passes the token to the next node (say node j) with probability P_{ij}^t. Eventually, the capacity share received by each data node satisfies (4.1). In order to let the proposed scheme works properly, the key point is to let all the data nodes have the same and accurate network information. In the WLAN, where all nodes can hear each other, to maintain such up-to-date information is not a difficult task. When a new data node wants to join the WLAN, it first broadcasts a JOIN message, announcing its class s. Upon receiving this JOIN message, all the existing data nodes re-calculate $\pi_t(i)$ and P_{ij}^t accordingly. One of the existing data node (e.g., the node currently holds the data token) sends a JOIN-ACK message, including the updated network information, to the new data node. The new data node then calculates its $\pi_t(i)$ and P_{ij}^t accordingly. Similarly, when a data node leaves the WLAN, it also broadcasts a LEAVE message. When the current data token holder leaves the WLAN, it passes the token to another node before departure. All the data nodes update their information accordingly. To avoid the potential collision between the JOIN/JOIN-ACK/LEAVE message transmission and the voice/data packet transmission, we let the data node wait for the channel being idle for a short period T_4 ($< T_3$) before sending JOIN/JOIN-ACK/LEAVE message. Thus, JOIN/JOIN-ACK/LEAVE messages have the highest priority to be sent.

It is possible that the proposed scheme consumes more power than contention-based schemes (e.g., IEEE 802.11) when the traffic load is low since the nodes still consume some power for token transmissions even if they have no packet to deliver. However, the proposed scheme becomes more and more power efficient than contention-based schemes when the traffic load becomes high because the power waste due to collisions in contention-based schemes will not occur in the proposed scheme.

4.2.3   Token Initialization and Recovery of Lost Tokens

The data token is initialized by the first data node (say node A) which joins a WLAN. After sending the JOIN message, node A waits for the JOIN-ACK message. If no response is received after a certain time period (i.e., the time needed for a JOIN-ACK message reception), node A considers itself as the first data node in the WLAN, and generates a data token. For any voice node in the WLAN, it first contends for the channel to transmit its first voice packet. After that, it monitors the channel, if no activity of other voice nodes is detected upon the next voice packet arrival, it will consider itself as the first voice node in the WLAN. It generates a voice token and passes it to itself. Once the activity of another voice node (say node B) is detected, it passes the voice token to node B.

The tokens may be lost due to the unreliable wireless channel. For voice nodes, two scenarios can happen. The first scenario is that a node (say node A) still receives voice token although it is in the off state. This occurs if the announcement message sent by node A (when it changed from on to off) was not correctly received by the node which sends the voice token to node A. In this case, node A resends the announcement message and passes the token to the next voice node. The second scenario is that a node in the on state does not correctly receive the voice token. We let the node (say node B) which passes the token to the next token holder (say node C) monitor the activity of node C. If the activity of node C is not observed (i.e., the token passed to node C is lost), node B resends the token. After several consecutive failed retransmissions, node B passes the token to the next voice node. When the number of backlogged packets of node C is more than one, node C will re-contend for the channel. Similarly, for data nodes, if the current token holder (say node D) cannot pass the token to the next token holder (say node E) after several consecutive transmissions, node D will pass the data token to another node (say node F). Next time when node D chooses node F as the next data token holder, it will replace node F with node E and pass the data token to node E so that the channel access opportunity of each data node remains unchanged.

The reasons that we choose the probabilistic token passing procedure for data traffic instead of deterministic token passing ring are as following. First, in a WLAN with nodes dynamically joining or leaving the WLAN, a deterministic token passing ring needs to be timely changed according to network dynamics. To maintain such a dynamic token passing ring among all the data nodes is not easy. In contrast, the probabilistic token passing procedure does not need to maintain such a token passing ring; Second, as aforementioned, with the probabilistic token passing procedure, it is relatively easy to recover a lost token and, at the same time, to keep the channel access opportunity of each data node unchanged. However, when a token is passed following a deterministic ring, it is difficult to do so.

4.3 Performance Analysis

To make the analysis tractable, we make the following assumptions: (a) The voice traffic follows the on/off model. The packet arrivals at each data node follow a Poisson process; (b) There is no packet loss in radio transmission and no node failure; (c) Each node has the same transmission rate. All the voice (or data) packets have the same size. When a data node gets the token, it transmits one data packet and then passes the token to the next node.

4.3.1 Voice Traffic Performance Analysis

4.3.1.1 The Channel Time Occupancy Fraction of Voice Traffic

Given N_v voice source nodes, the fraction of channel time used by voice traffic, denoted by ψ, can be derived as follows. The traffic from each voice node follows the on/off model, and the durations of the on and off states are exponentially distributed with mean values $1/\alpha$ and $1/\beta$, respectively. Hence, at any time instant, each voice node is at the on state with probability $\beta/(\alpha+\beta)$. During each voice packet inter-arrival duration (denoted by I_o), each voice node which is at the on state generates one voice packet. Thus, the average channel time used by voice traffic during I_o is given by

$$\overline{T} = \sum_{i=1}^{N_v} \binom{N_v}{i} \left(\frac{\beta}{\alpha+\beta} \right)^i \left(\frac{\alpha}{\alpha+\beta} \right)^{N_v-i} \cdot i \cdot T_{voice}, \quad 0 \leq i \leq N_v \qquad (4.3)$$

where T_{voice} is the voice packet transmission time. Then the fraction ψ is given by $\psi = \frac{\bar{T}}{\bar{T}_o}$. On the other hand, given ψ, we can determine the maximum number of voice nodes N_v that can be admitted to the network. This result facilitates call admission control of voice traffic when we need to guarantee data traffic a fraction of the channel time.

4.3.1.2 Voice Delay

The delay is defined as the time period from the moment that a packet arrives at a node to the moment that the packet is successfully transmitted from the node. The voice node at the on state can be modeled by a D/G/1 queue model, where the packets arrive at a constant rate, and the service time is the voice token recurrence time, which is defined as the time duration between two consecutive token passing time instants of a node. As discussed in Sect. 4.2, the token passing sequence keeps track of the packet arrival orders of different nodes, and voice packets arrive periodically, so the variance of the voice token recurrence time is expected to be small (which is verified by simulations). Thus, the queue can be approximated by a D/D/1 queue, where the queueing delay is small, and is bounded by the packet inter-arrival time.

4.3.1.3 Collision Probability of Voice Nodes from the off State to the on State

Here, we are interested in the collision probability in the worst case, where the collisions of voice nodes from the off state to the on state are most likely to happen. Since a data packet needs a long time duration to transmit (compared with a voice packet and a token frame), it is more likely that collisions of voice nodes occur after data packet transmissions. Thus, the worst case occurs when data traffic are saturated (i.e., the data source nodes always have a packet to transmit). At any time instant, a voice node is at the off state with probability $\alpha/(\alpha+\beta)$. Given that a voice node is at the off state, the conditional probability that a transition to the on state happens within duration t is given by $1 - e^{-\beta t}$. In our case, t can be a data packet transmission time T_{data}, or a voice packet transmission time T_{voice}. Here, we do not take token transmission time into account. In the worst case, the token frame transmissions rarely happen since all the data tokens are piggybacked over data packets, and the voice token is transmitted only when a voice source node is from the on state to the off state. Besides, since the token transmission time is very short, the fraction of channel time used for token frame transmissions is negligible. Given N_v voice nodes in the network, the probability that a collision happens after a data packet transmission, denoted by P_c^{data}, is given by (4.4).

$$P_c^{data} = \sum_{i=1}^{N_v} \binom{N_v}{i} \left(\frac{\alpha}{\alpha+\beta}\right)^i \left(\frac{\beta}{\alpha+\beta}\right)^{N_v-i} \cdot \left[1 - \left(e^{-\beta T_{data}}\right)^i\right.$$

$$\left. -i \cdot \left(e^{-\beta T_{data}}\right)^{i-1} \cdot \left(1 - e^{-\beta T_{data}}\right)\right], \quad 0 \le i \le N_v. \tag{4.4}$$

$$T_r^i = \psi \cdot T_r^i + T_{data} \cdot \rho_i + T_{token} \cdot (1 - \rho_i)$$

$$+ \sum_{j=1, j \ne i}^{N_d} \frac{\pi_t(j)}{\pi_t(i)} \cdot [T_{data} \cdot \rho_j + T_{token} \cdot (1 - \rho_j)], \quad \rho_i \le 1, \rho_j \le 1, i = 1, 2, \ldots, N_d. \tag{4.5}$$

Similarly, by replacing T_{data} with T_{voice} in (4.4), we get P_c^{voice}, the probability that a collision happens after a voice packet transmission. For any packet transmission, the probability that it is a voice packet transmission is

$$\frac{\psi/T_{voice}}{\psi/T_{voice} + (1-\psi)/T_{data}}.$$

Then the collision probability in the worst case, P_c^w, is given by

$$P_c^w = \frac{\psi/T_{voice}}{\psi/T_{voice} + (1-\psi)/T_{data}} \cdot P_c^{voice} + \frac{(1-\psi)/T_{data}}{\psi/T_{voice} + (1-\psi)/T_{data}} \cdot P_c^{data}.$$

4.3.2   Data Traffic Performance Analysis

4.3.2.1   Data Throughput

The system throughput is defined as the ratio of the time used for data packet transmission to the total channel time. Consider a WLAN with N_d data source nodes and N_v voice source nodes. For the ith data node, let λ_i denote the average packet arrival rate. To calculate the system throughput, we first derive the average token recurrence time of node i, denoted by T_r^i. As discussed, for N_v voice nodes, the voice traffic occupy a constant fraction ψ of the channel time. Since the arrival time of the first packet of a talk burst at each voice node is random, we assume that the channel time occupied by all the voice traffic is uniformly distributed over the total channel time. Thus, during T_r^i, the voice traffic occupies $\psi \cdot T_r^i$ channel time on average. The average service rate of node i, μ_i, is simply the reciprocal of its token recurrence time. The queue utilization ratio at node i is denoted by $\rho_i = \frac{\lambda_i}{\mu_i} = \lambda_i \cdot T_r^i$. Node i is empty (with no packet to send) with probability $1 - \rho_i$. When a data token holder has a packet to transmit, it takes T_{data} to transmit; otherwise, it takes T_{token} to pass the token to the next node. The steady-state probability $\pi_t(i)$ given in (4.1) reflects the frequency that node i holds the data token. In a long term, in the time

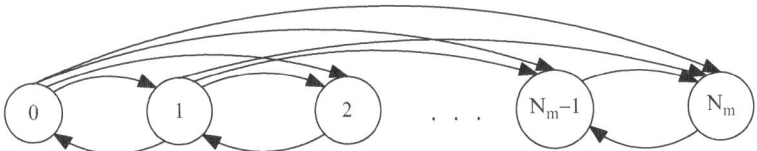

Fig. 4.2   The state transition diagram of the imbedded Markov chain

interval of T_r^i (during which node i holds the data token once), on average, node j holds the data token $\frac{\pi_t(j)}{\pi_t(i)}$ times. Thus, we have (4.5). Solving (4.5), we get the token recurrence time T_r^i ($i = 1, 2, \ldots, N_d$). Then the system data throughput is given by

$$T_s^w = (1 - \psi) \cdot \frac{\sum_{i=1}^{N_d} \pi_t(i) \cdot T_{data\_payload} \cdot \rho_i}{\sum_{i=1}^{N_d} \pi_t(i) \cdot (T_{data} \cdot \rho_i + T_{token} \cdot (1 - \rho_i))}, \ \rho_i \leq 1$$

where $T_{data\_payload}$ is the time to transmit the payload of a data packet. When $\rho_i = 1$ for all the data nodes, the system is in an overload condition.

4.3.2.2   Data Packet Delay

With the Poisson arrival assumption, the packet arrival and departure at each data node can be modeled by an M/G/1 queue. To determine the average delay, we first need to obtain the average queue length at each data node. According to [49] (p. 175), for an M/G/1 queue, arrivals, departures, and random observers all see the same distribution of the number of customers in the system. Thus, we conclude that the average queue length at an arbitrary time is equal to the average queue length at any packet departure instant. We consider an imbedded Markov chain in which the state transitions occur at the packet departure instants of a tagged node. We define the state of this imbedded Markov chain to be the number of packets left behind by the departing packet. For simplicity of analysis, we assume that the probability of a node having more than N_m (N_m is chosen to be a large number) packets in the queue is negligible. The state transition diagram of the imbedded Markov chain is shown in Fig. 4.2. The average queue length at the tagged node is given by $L = \sum_{k=0}^{N_m} k p(k)$, where $p(k)$ is the steady-state probability of state k. To find the steady-state probability vector of this Markov chain, we should first obtain the state transition probability from any state i to j, denoted by $pr(i, j)$.

The derivation of $pr(i, j)$ for the case $i = 0$ is more complex than for the case $i > 0$. We first consider the case $i > 0$, i.e., the tagged node is backlogged (when a packet leaves the node, there is at least one packet left in the queue). $pr(i, j)(i > 0)$ is the probability that, during a token recurrence time of a backlogged data node, $j - i + 1$ packets arrive at that node. Obviously, for all $j \leq i - 2$, $pr(i, j) = 0$. Denote the average traffic arrival rate at the tagged node as λ, we have

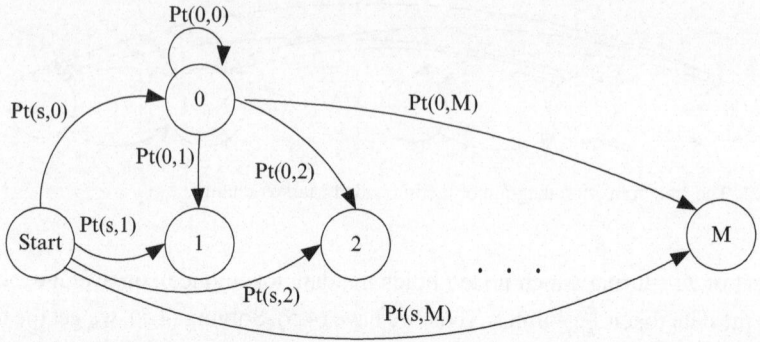

Fig. 4.3 The state transition diagram of the sampled process

$$pr(i,j) = \int_0^\infty \frac{(\lambda x)^{j-i+1}}{(j-i+1)!} e^{-\lambda x} b(x) dx, \quad i > 0 \tag{4.6}$$

where $b(x)$ is the PDF (probability density function) of the token recurrence time at the tagged node, which is backlogged. In order to obtain $pr(i,j)$, we need to know the distribution of this token recurrence time. Since this distribution is difficult to obtain directly, we use the Laplace transform $B^*(s)$ for $b(x)$ defined as $B^*(s) = \int_0^\infty e^{-sx} b(x) dx$ (The derivation of $B^*(s)$ is presented in Sect. 4.3.2.3). We define random variable v as the number of packet arrivals during a token recurrence time of a backlogged data node, and the z-transform of the PMF (probability mass function) of this random variable is given by $V(z) = \sum_{k=0}^\infty P[v = k] z^k$. The relationship between $V(z)$ and $B^*(s)$ is as follows [49] (p. 184),

$$V(z) = B^*(\lambda - \lambda z). \tag{4.7}$$

Then $pr(i,j)(i > 0)$ is given by

$$pr(i,j) = \frac{1}{(j-i+1)!} \cdot \frac{d^{j-i+1}}{dz^{j-i+1}} V(z)|_{z=0}.$$

For the derivation of $pr(i,j)$ with $i = 0$ (i.e., when a packet leaves the tagged node, the node has an empty queue), we consider two scenarios. The first scenario is that, when the tagged node gets a new chance to transmit (we denote this time instant as t_s, which is the time instant that the node gets the data token again and the channel is available for the data node to transmit), the queue is still empty. In this case, it will pass the token immediately to another node. The other scenario is that when the tagged node gets a new chance to transmit, it has at least one packet in the queue. In this case, it will transmit the data packet and piggyback the token over the data packet. Once state 0 in Fig. 4.2 occurs, we sample the number of packets at the tagged node, denoted by k, at the time instants t_s, and stop sampling when $k > 0$. Figure 4.3 shows the state transition diagram of the sampled process. There is one starting state (the same as state 0 in Fig. 4.2), one transient state ($k = 0$), and N_m absorbing states ($k = 1, 2, \ldots, N_m$), with state transition probabilities $Pt(\cdot, \cdot)$. The derivation of these state transition probabilities is similar to that of $pr(i, j)(i > 0)$. Note that different time intervals are involved. The time interval from the starting state to state k ($k = 0, 1, \ldots, N_m$), denoted by t_1, is the time period from the moment that the last packet in the queue leaves the tagged node to the moment that the node gets a new chance to transmit. The time interval from the transient state to an absorbing state, denoted by t_2, is token recurrence time of an empty node. Both t_1 and t_2 are random variables and the Laplace transforms of their PDFs are represented by $H_1^*(s)$ and $H_2^*(s)$, respectively (The derivations of $H_1^*(s)$ and $H_2^*(s)$ are given in Sect. 4.3.2.3). Let v_1 and v_2 denote the number of packet arrivals during t_1 and t_2, respectively, and $V_1(z)$ and $V_2(z)$ the z-transform of the PMFs of v_1 and v_2, respectively. Similar to (4.7), we have $V_1(z) = H_1^*(\lambda - \lambda z)$ and $V_2(z) = H_2^*(\lambda - \lambda z)$.

The transition probability from the starting state to state k ($k = 0, 1, \ldots, N_m$), denoted by $Pt(s, k)$, is given by

$$Pt(s, k) = \frac{1}{k!} \cdot \frac{d^k}{dz^k} V_1(z)|_{z=0}.$$

Also, the transition probability from the transient state ($k = 0$) to state k ($k = 0, 1, \ldots, N_m$), denoted by $Pt(0, k)$, is given by

$$Pt(0, k) = \frac{1}{k!} \cdot \frac{d^k}{dz^k} V_2(z)|_{z=0}.$$

Denote $Pa(k)$ the probability that the sampled process enters absorbing state k ($k = 1, 2, \ldots, N_m$). Given the state transition probabilities, it is straightforward to get $Pa(k)$ as follows

$$Pa(k) = Pt(s, k) + \frac{Pt(s, 0)Pt(0, k)}{1 - Pt(0, 0)}, \quad k = 1, 2, \ldots, N_m.$$

$Pa(k)$ is actually the conditional probability that the tagged node (which finds itself empty when the latest packet departs) finds $k(> 0)$ packets in the queue when it gets a new chance to transmit under the condition that, when it gets the new chance to transmit, it finds at least one packet present. Then we have

$$pr(0, j) = \sum_{k=0}^{j} Pa(k+1) \frac{e^{-\lambda T_{data}} (\lambda T_{data})^{j-k}}{(j-k)!}, \quad j = 0, 1, \ldots, N_m. \quad (4.8)$$

We take $pr(0, 0)$ and $pr(0, 1)$ as examples to explain (4.8). $pr(0, 0)$ represents the possibility that, when the tagged node gets a chance to transmit, it finds one packet in the queue (with probability $Pa(1)$), and there is no packet arrival (with probability $e^{-\lambda T_{data}}$) during the packet transmission time. For $pr(0, 1)$, there are two cases: One case is that, when the tagged node gets a chance to transmit, it finds one

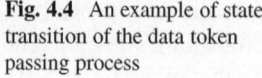

Fig. 4.4 An example of state transition of the data token passing process

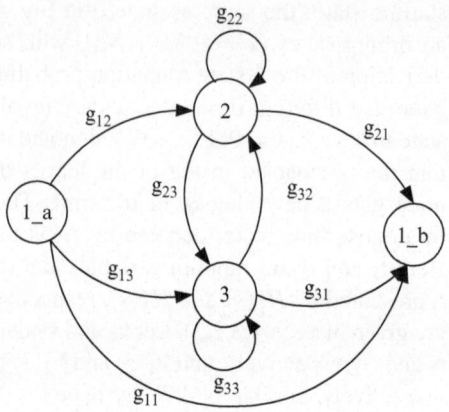

packet in the queue (with probability $Pa(1)$) and, during the packet transmission time, one new packet arrives (with probability $e^{-\lambda T_{data}}\lambda T_{data}$); The other case is that it finds two packets in the queue (with probability $Pa(2)$), and no packet arrives (with probability $e^{-\lambda T_{data}}$) during the packet transmission time.

With $pr(i,j)$, we can obtain the steady-state probabilities $p(k)$ ($k = 0,1,\ldots,N_m$) of the chain shown in Fig. 4.2. Then we have the average queue length L. According to the Little's law, the average delay at the tagged node is given by $D = \frac{L}{\lambda}$.

4.3.2.3 The Derivation of $B^*(s), H_1^*(s)$, and $H_2^*(s)$

$B^*(s)$ is the Laplace transform of the PDF of a backlogged data node's token recurrence time. As discussed, the data token passing process in the proposed scheme can be modeled by a Markov chain. Each transition from state i to j represents that the token is passed from node i to node j. In order to get $B^*(s)$, we re-draw this Markov chain as follows.

For simplicity of presentation, we take a simple example with three data nodes in the WLAN and take node 1 as the tagged node. We split state 1 into two states: 1_a and 1_b, as shown in Fig. 4.4. State 1_a is the starting point, representing that node 1 is passing the data token to others; and state 1_b is the ending point, representing that node 1 is receiving the data token from others. Assuming node 1 is backlogged, the time duration taken from state 1_a to state 1_b is actually the token recurrence time of a backlogged data node, whose PDF has Laplace transform $B^*(s)$.

To get $B^*(s)$, a method similar to that used in [101] is applied here. The Markov chain shown in Fig. 4.4 can be treated as a signal flow graph [71]. State 1_a is the source of the signal and state 1_b is the sink. All the other nodes are signal repeaters. Each branch (state transition) is associated with a branch transmittance. The signals travel along the branches and are modified by the corresponding branch transmittances. A repeater combines all the incoming signals and sends the outgoing signal along all the branches diverging from that repeater. The signal transfer function from the source to the sink can be obtained by application of the Mason's rule or other flow reduction methods [71]. It has been found out that, if the branch transmittances are properly defined, the probability generating function of total transition time from the source to the sink can be obtained from the signal transfer function [101]. The probability generating function is actually the z-transform of the PMF of the total transition time from the source to the sink. Since Laplace transform has many of the same properties as z-transform, by properly defining the branch transmittances, we can get the Laplace transform of the PDF of the total transition time from the source to the sink.

Next, we discuss how to define the branch transmittances. In the following, instead of using the transition probability, we use branch transmittance, denoted by g_{ij}, to associate with the state transition from i to j. The transmittance is defined as $g_{ij} = P_{ij}^t \cdot e^{-st}$, where P_{ij}^t is the transition probability from state i to j (i.e., data token passing probability from node i to j), t is the transition time from state i to j (i.e., the time duration that the process remains in state i before transiting to state j), and s is a dummy variable. For a specific state transition, the transition time may not be a constant. For example, considering the transition from state 2 to 3, when node 2 has packets to send, it takes T_{data} to transit to state 3; otherwise, it takes T_{token} to transit to state 3. Considering the presence of voice traffic, the data token holder has to wait for the voice traffic to be transmitted first. During the voice packet inter-arrival time I_o, on average, the ith data node holds the token I_o/T_r^i times, where T_r^i is given in (4.5). So for each data transmission, the time delayed by voice traffic is given by $T_{delay} = \dfrac{\overline{T}}{I_o \sum_{i=1}^{N_d} \frac{1}{T_r^i}}$, where $\overline{T}$ is given in (4.3). So the term e^{-st} is re-written as $po_2 \cdot e^{-s(T_{token}+T_{delay})} + (1 - po_2) \cdot e^{-s(T_{data}+T_{delay})}$, where po_2 is the probability that node 2 has no packet to send. As discussed in Sect. 4.3, we have $po_2 = 1 - p_2$. Notice that node 1 is assumed to be backlogged, so the transition time from state 1_a to other states is a constant (i.e., $T_{data} + T_{delay}$). From the above discussion, we have

$$g_{ij} = \begin{cases} P_{ij}^t e^{-s(T_{data}+T_{delay})}, & i = 1 \\ P_{ij}^t \cdot \left[po_j \cdot e^{-s(T_{token}+T_{delay})} + (1 - po_j) \cdot e^{-s(T_{data}+T_{delay})} \right], & i \neq 1. \end{cases}$$

According to [71], the signal transfer function from the source 1_a to the sink 1_b, denoted by $STF(1\_a, 1\_b)$, is given by

$$STF(1\_a, 1\_b) = g_{11} + \frac{g_{13} \cdot g_{31}}{1 - g_{33}} + \frac{\left(g_{12} + \frac{g_{13} \cdot g_{32}}{1 - g_{33}}\right) \cdot \left(g_{21} + \frac{g_{23} \cdot g_{31}}{1 - g_{33}}\right)}{1 - g_{22} - \frac{g_{32} \cdot g_{23}}{1 - g_{33}}}. \tag{4.9}$$

Based on the conclusion drawn in [101], $B^*(s)$ is simply equal to $STF(1\_a, 1\_b)$.

Recall that $H_1^*(s)$ is the Laplace transform of the PDF of the random variable t_1, which is the time period from the moment that the last packet in the queue leaves a data node to the moment that the node gets a new chance to transmit. Let random variable x denote the token recurrence time of a backlogged data node, whose PDF has Laplace transform $B^*(s)$. By the definition of t_1 and x, we have $t_1 = x - T_{data}$. According to the property of Laplace transform, we have $H_1^*(s) = B^*(s)e^{-sT_{data}}$.

$H_2^*(s)$ is the Laplace transform of the PDF of the random variable t_2, which is the token recurrence time of an empty data node. The derivation of $H_2^*(s)$ is similar to that of $B^*(s)$. The only difference is in the calculation of g_{ij}. By the definition of t_2, node 1 (corresponding to state 1_a in Fig. 4.4) is empty now. So the transition time taken from state 1_a to any other state is $T_{token} + T_{delay}$. Similarly, the term e^{-st} is re-written as $po_j \cdot e^{-s(T_{token}+T_{delay})} + (1 - po_j) \cdot e^{-s(T_{data}+T_{delay})}$ ($j = 2, 3$) for the transitions from node 2 and node 3. Thus, for $H_2^*(s)$, we have

$$g_{ij} = \begin{cases} P_{ij}^t e^{-s(T_{token}+T_{delay})}, \ i = 1 \\ P_{ij}^t \cdot [po_j \cdot e^{-s(T_{token}+T_{delay})} + (1 - po_j) \cdot e^{-s(T_{data}+T_{delay})}], \ i \neq 1. \end{cases}$$

Applying g_{ij} to (4.9), we get $H_2^*(s)$.

4.4 Numerical Results and Performance Evaluation

In this section, we validate our analysis and evaluate the performance of the proposed scheme by extensive simulations. For voice traffic, we compare delay performance with IEEE 802.11e. We choose the GSM 6.10 codec as the voice source as an example. For data traffic, we compare the performance of channel utilization with IEEE 802.11 DCF (which is a contention-based scheme) and the centralized polling scheme [1]. For the contention-based scheme, since the RTS/CTS mechanism can improve the performance compared with the basic access scheme [11], we adopt the RTS/CTS mechanism in our simulation. In the polling scheme, a central controller polls each node (based on its scheduling policy) by broadcasting a polling frame. Upon being polled, a node is granted a transmission opportunity to transmit its packets. If the polled node has no packet to send, the central controller polls next node immediately. For fair comparison, the proposed scheme and the polling scheme have the same scheduling policy, each polled node (or data token holder) is granted the same channel time, and the data token size is chosen to be the same as the polling frame size. The simulation parameters are given in Table 4.1, where the channel rate is to transmit voice/data packets, and the basic rate is to transmit RTS, CTS, polling frames and the token. The simulation is done in Matlab. Each data node uses a random generator to randomly choose the next token holder based on the transition probabilities. In each run, we simulate 50 s of the channel time (except those scenarios which have a specific simulation end time). Each of the simulation results represents an average of 10 independent runs.

Table 4.1 System
parameters used in simulation
and analysis of the
token-based MAC scheme

Parameter	Value
Slot time	20 μs
T_1/AIFS[AC_data]	60 μs
T_2/AIFS[AC_voice]	40 μs
T_3	20 μs
T_4/SIFS	10 μs
CW_{min}[voice]	15
CW_{max}[voice]	63
CW_{min}[data]	31
CW_{max}[data]	1,023
PHY preamble	192 μs
RTS frame size	20 bytes
CTS frame size	14 bytes
Polling/token frame size	36 bytes
Data packet size	1,000 bytes
Voice packet size	107 bytes
Channel rate	11 Mbps
Basic rate	2 Mbps
$1/\alpha$	352 ms
$1/\beta$	650 ms
I_o	20 ms

Table 4.2 The channel time fraction occupied by voice traffic

The number of voice source nodes	20	30	40	50	60
Simulation results (%)	11	17	23	28	35
Analytical results (%)	11	16	21	26	32

4.4.1 Voice Traffic Analysis Accuracy

Table 4.2 shows the fraction of the channel time occupied by voice traffic with different number of voice source nodes (N_v). It can be seen that the analytical results match closely with the simulation results. For the collision probability (P_c^w), the simulation results demonstrate that it is small, even when N_v is large. For N_v equals to 110, 120 and 130, the analytical results of P_c^w are 0.14, 0.15 and 0.15 %, respectively, while the simulations results are 0.14, 0.15, and 0.16 %, respectively. For delay performance, we consider an integrated voice/data scenario with $N_v = 50$. Figure 4.5 compares the delays of the proposed scheme and IEEE 802.11e with different number of data source nodes. It can be seen that the average voice delay increases greatly with the increase of data source nodes when using IEEE 802.11e, but remains unchanged in the proposed scheme. The reason is that our scheme provides guaranteed priority to voice nodes, thus the voice performance is not affected by the data traffic. We also observe that the voice delay in our scheme is very low (around 1 ms), which verifies the conclusion drawn in our analysis.

Fig. 4.5 The voice packet
delay versus the number of
data source nodes with
$N_v = 50$

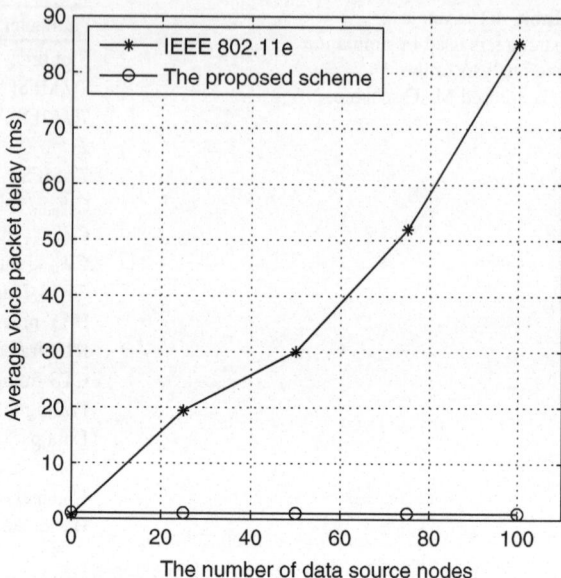

4.4.2 Proportional Class Differentiation of Data Traffic

Since the performance of proportional class differentiation to data traffic is not
affected by the voice traffic, for simplicity, we consider a WLAN with 20 data
source nodes in the absence of voice nodes. First, we vary the class differentiation
requirements to see if the proposed scheme can provide quantitative class differentiation based on an arbitrary requirement. We consider three scenarios with different
classes and required class differentiation ratios. In the first scenario, the nodes are
classified into two classes, each having 10 nodes. The required differentiation ratio
is 1 : 2. In the second scenario, there are three classes with a desired differentiation
ratio 1 : 1.5 : 3. The number of nodes in class 1, 2 and 3 are 5, 5 and 10, respectively.
In the third scenario, the nodes are grouped into four classes, each having five nodes.
The differentiation ratio is 1 : 0.5 : 2 : 3. The throughput of each node in the three
scenarios are shown in Table 4.3. It is clear that the nodes of different classes achieve
different throughputs. The ratios are very close to the requirements. The nodes in
the same class achieve almost the same throughput as expected. The simulation
results demonstrate that the proposed scheme can effectively provide proportional
class differentiation based on a specific differentiation requirement.

Note that in all the above three scenarios, the number of nodes in the WLAN
remains unchanged during the whole simulation time. Next, we vary the number of
nodes (i.e., some nodes join or leave the WLAN) to verify that the proposed scheme
is adaptive to the network dynamic and provides consistent class differentiation.
Consider three classes, with required class differentiation ratio 1 : 2 : 3. At the
beginning of the simulation (i.e., $t = 0$), there are five nodes in each class. At $t = 5$ s,

Table 4.3 The throughput achieved by each data node in the three scenarios with different classes

	Scenario 1		Scenario 2		Scenario 3	
Node no.	Class no.	Throughput (Mbps)	Class no.	Throughput (Mbps)	Class no.	Throughput (Mbps)
1	1	0.2819	1	0.2027	1	0.2626
2	1	0.2826	1	0.2047	1	0.2666
3	1	0.2750	1	0.2043	1	0.2620
4	1	0.2826	1	0.1993	1	0.2602
5	1	0.2822	1	0.2020	1	0.2634
6	1	0.2859	2	0.2985	2	0.1318
7	1	0.2852	2	0.3032	2	0.1315
8	1	0.2836	2	0.2996	2	0.1315
9	1	0.2837	2	0.2963	2	0.1358
10	1	0.2861	2	0.2980	2	0.1305
11	2	0.5643	3	0.6066	3	0.5234
12	2	0.5776	3	0.5973	3	0.5278
13	2	0.5685	3	0.5897	3	0.5235
14	2	0.5782	3	0.6095	3	0.5274
15	2	0.5710	3	0.6114	3	0.5248
16	2	0.5727	3	0.6081	4	0.7900
17	2	0.5752	3	0.5945	4	0.8001
18	2	0.5724	3	0.6048	4	0.7910
19	2	0.5679	3	0.6052	4	0.7857
20	2	0.5754	3	0.6063	4	0.8023

a new node of class 1 joins the WLAN and, at $t = 10$ s, a node in class 3 leaves the WLAN and, at $t = 15$ s, a new node of class 2 joins the WLAN. We obtain the throughput of each node, shown in Fig. 4.6. The throughput of each node is reduced from $t = 5$ s because of the new node arrival, and increases from $t = 10$ s because of the node departure, and reduced again from $t = 15$ s because a new node joins the WLAN. Note that the throughput ratios among the three classes remain very close to the constant (as $1 : 2 : 3$) during the whole simulation time. Although the actual throughput of each node changes due to the network dynamics, the proposed scheme maintains a consistent class differentiation ratio among different classes.

4.4.3 Data Throughput and Delay Analysis Accuracy

Consider the WLAN with two classes, each having 15 data source nodes. The number of voice source nodes is 20. The required class differentiation ratio is $1 : 2$. Figure 4.7 shows the aggregate throughput and the throughput achieved by each class, with different system traffic loads. For delay performance, consider three classes with a required class differentiation ratio $1 : 2 : 3$. Figure 4.8 shows the

Fig. 4.6 The throughput achieved by a node in each class

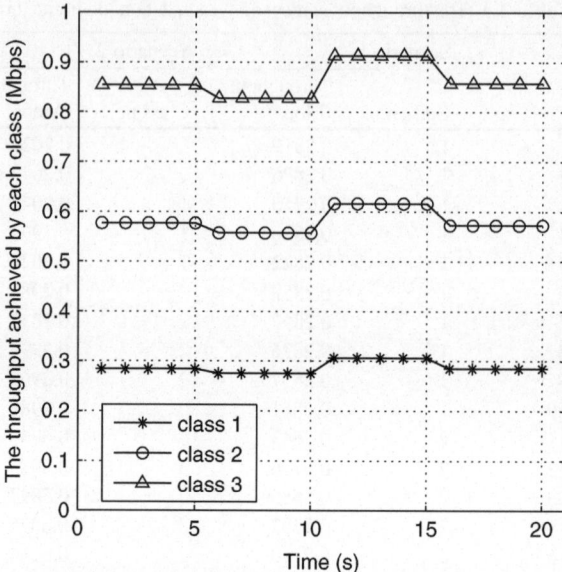

Fig. 4.7 The throughput versus the system traffic load with $N_v = 20$

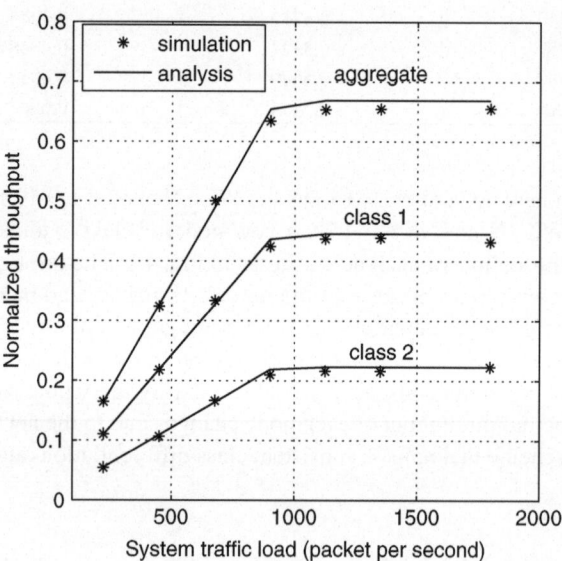

average delay of the three classes versus the system traffic load. Obviously, when the traffic load becomes high, the delay suffered by the node in each class increases. From Figs. 4.7 and 4.8, it is clear that the analytical and simulation results of the throughput and delay agree with each other very well.

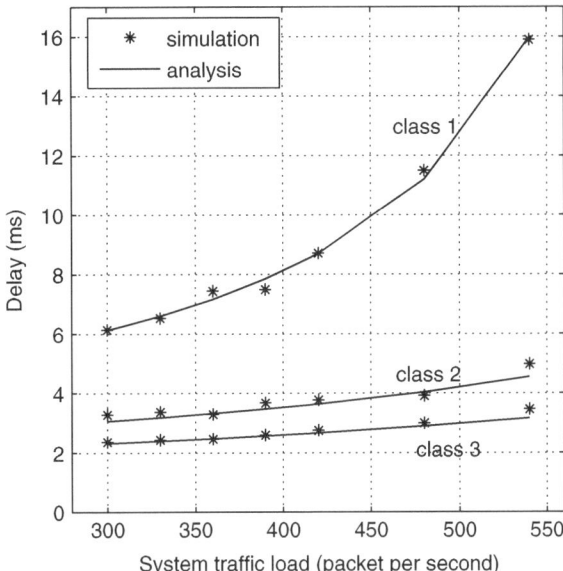

Fig. 4.8 The average packet delay versus the system traffic load with $N_v = 20$

4.4.4 Channel Utilization

Channel utilization is represented by the ratio of the achieved system throughput to the channel rate. From the simulations, we find that the channel utilization performance is not sensitive to class differentiation. For simplicity, here we consider a homogeneous WLAN with a single class. We fix the number of the data source nodes to be 20 in the WLAN, and vary the traffic arrival rate. Figure 4.9 compares the channel utilization of IEEE 802.11 DCF, the polling scheme, and the proposed scheme over a perfect channel. It is clear that the proposed scheme achieves much higher channel utilization than IEEE 802.11 DCF, when the traffic load becomes high. With an increase of traffic load, collisions occur more frequently with the contention-based scheme. By avoiding those collisions, the proposed scheme utilizes the channel more efficiently. Compared with the centralized polling scheme, the proposed scheme also achieves higher utilization, because it reduces the overhead incurred by the polling frames. The channel utilization of the proposed scheme over an unreliable channel is also shown in Fig. 4.9, where the packet error probability $P_e = 5\%$. The impact of the channel is negligible when the traffic load is low, but results in an approximate 5% reduction in the channel utilization. It is observed that the proposed scheme still outperforms the other two schemes, taking into account of the possible token loss.

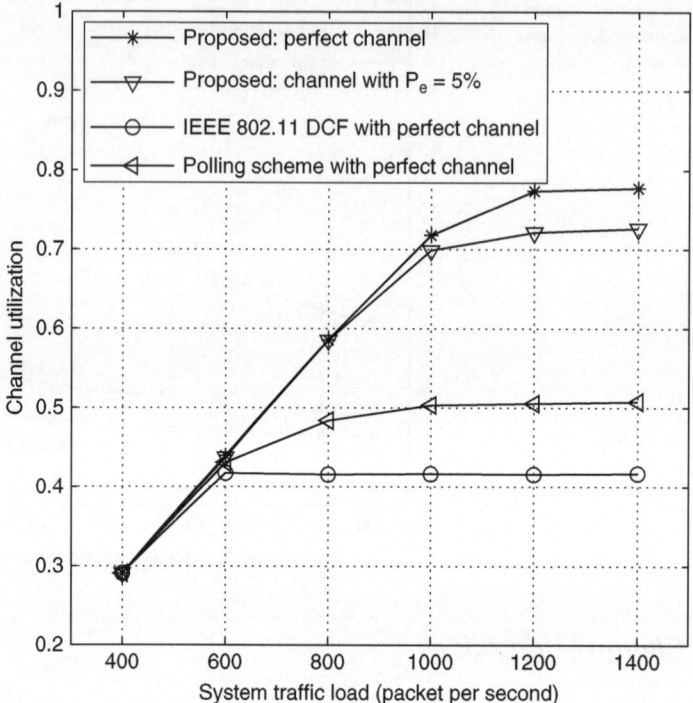

Fig. 4.9 The channel utilization versus the system traffic load

4.5 Summary

In this chapter, we propose a novel token-based scheduling scheme for an ad hoc
WLAN that supports both voice and data traffic. The proposed scheme can provide
guaranteed priority access to voice traffic and, at the same time, provide precise and
quantitative service differentiation for data traffic, which provides great flexibility
and facility to the network service provider for service class management. The
advantages of the proposed scheme are summarized as follows:

- Most of the conventional contention-based MAC schemes (e.g., IEEE 802.11e)
  provide statistical priority to voice traffic, in which the performance of voice
  service degrades with an increase of the data traffic load. On the contrary,
  the proposed scheme provides guaranteed priority to voice traffic, thus the voice
  performance is not affected by the data traffic load.
- Compared with the conventional contention-based MAC schemes, the service
  class differentiation can be achieved quantitatively in the proposed scheme.
  Each class can get exactly the desired portion of the channel capacity. Such a
  class differentiation is difficult to achieve by adjusting the contention windows
  (or inter-frame spaces) in contention-based schemes.

- The contention-based MAC schemes are subject to collisions. This nature renders those schemes inefficient channel utilization. By passing the token among the nodes in the WLAN, the proposed scheme eliminates collisions which occur in contention-based schemes; therefore, it achieves higher resource utilization, especially when the traffic load is high.

Chapter 5
Dual Busy-Tone MAC for Wireless Ad Hoc Networks

In the preceding chapters, a single-hop wireless network (i.e., WLAN) is considered. In this chapter, we consider a multi-hop wireless ad hoc network. As discussed in Chap. 2, unlike single-hop WLANs, the multi-hop wireless network presents more challenges to the QoS provisioning. The hidden terminals bring more collisions. The exposed terminals lead to inefficient channel utilization. The locations of the contending flows can greatly affect the channel access opportunity of each flow, resulting in serious unfairness and priority reversal problems. We propose an effective MAC scheme to address all these problems. Our proposed MAC scheme utilizes two narrow-band busy-tone channels and one information channel. Similar to all other busy-tone schemes, extra hardware cost is incurred to implement the busy-tone channels. However, as mentioned in [33], the wireless transceiver architecture proposed in [32] can help to set up the busy-tone channels with low hardware cost.

5.1 Wireless Ad Hoc Network

A wireless ad hoc network is a collection of wireless mobile nodes that self-configure to form a network without the aid of any pre-existing infrastructure. There is no central controller (such as base stations in cellular systems) to organize the network and schedule the transmissions from the mobile nodes. In such a network, each node can be a source, a destination or a relay node for others. The source and destination pairs are arbitrary and they communicate via single or multi-hop wireless links.

Similar to the IEEE 802.11b standard and most of the research in ad hoc networks, direct sequence spreading spectrum (DSSS) is applied at the physical layer. Different from a direct sequence–code-division multiple access (DS-CDMA) cellular system [69], where each mobile node has a unique pseudo-noise (PN) code, in our system, one common code is used by all the mobile nodes to transmit data,

P. Wang and W. Zhuang, *Distributed Medium Access Control in Wireless Networks*,
SpringerBriefs in Computer Science, DOI 10.1007/978-1-4614-6602-4_5,
© The Author(s) 2013

making it impossible to have simultaneous transmissions within a receiver's vicinity. The reason of using one common code, instead of multiple PN codes, is due to the characteristics of ad hoc networks. In cellular networks, a base station is a central controller, collecting all the data transmission requests from mobile nodes through the uplink, and broadcasting the scheduling result to all the mobile nodes through the downlink. Therefore, simultaneous transmissions are easy to handle in such a system. However, in ad hoc networks, transmission pairs are arbitrary, and there is no central controller to schedule the transmission time of each node. To allow simultaneous transmissions within a receiver's vicinity, extra signaling has to be exchanged among different nodes to control the interference and coordinate the order of transmissions. It brings about much implementation complexity and a large signaling overhead to the system. By using a common code, extra transmissions within a receiver's vicinity are prohibited, so it is relatively easy to coordinate the transmissions of mobile nodes.

5.2   The Dual Busy-Tone MAC Scheme

In our proposed MAC scheme, the total channel bandwidth is divided into three parts with sufficient spectral separation (which is similar to DBTMA [33]): information channel, transmitter busy-tone (BTt) channel, and receiver busy-tone (BTr) channel. The difference of our scheme from DBTMA is that, by adjusting the receiver's sensitivity, we set the channels' carrier sense ranges such that the BTt channel's carrier sense range covers the two-hop neighborhood of the sensing node, while the BTr channel's carrier sense range covers the one-hop neighborhood of the sensing node.[1] The reason for such a setting is explained in the following subsections.

Similar to the IEEE 802.11e, voice and data traffic are assigned different AIFS values, i.e., AIFS[voice] < AIFS[data]. Before its contention, each contending node should wait for the two busy-tone channels idle for a duration of its AIFS. Each node also keeps a backoff timer, the initial value of which is randomly selected from its contention window. After the AIFS idle time of the two busy-tone channels, the node starts to send a busy tone in the BTt channel (instead of starting to count down its backoff timer, as in the IEEE 802.11e). The duration of the busy tone equals its

---

[1]All the nodes which are within the transmission range of a node (say node A) are one-hop neighbors of node A. All the nodes which are beyond the transmission range of node A but within two times the transmission range of node A are two-hop neighbors of node A. In a wireless network, the carrier sense range varies with the receiver's sensitivity. The feasibility of such a setting method can be found in [100]. For presentation clarity, we assume that, when a node is receiving a frame, only its one-hop neighbours' transmissions may corrupt its reception. In reality, a node's interference range may be larger than its transmission range, so the nodes beyond one-hop of a receiver may still be able to corrupt the reception. In this case, our scheme can still work if we adjust the BTt channel's carrier sense range to be the interference range plus the transmission range, and the BTr channel's carrier sense range to be the interference range.

backoff timer (in the unit of slot time). Upon the completion of its busy tone, the node senses the BTt channel again. If a busy BTt channel is sensed (i.e., another node is sending a busy tone), the node selects a new backoff timer (from its current contention window), and starts its busy tone after AIFS idle time again of both the busy-tone channels. If the BTt channel is idle:

- For the case of voice traffic, the voice transmitter sends its voice-DATA frame,[2] and simultaneously sends a busy tone in the BTt channel until the completion of the voice-DATA frame, for the purpose of protecting the voice-DATA frame from being corrupted by hidden terminals (to be discussed in Sect. 5.2.2). Upon reception of the voice-DATA frame, the receiver sends a busy tone in the BTr channel, which serves as an ACK.
- For the case of data traffic, the data transmitter sends an RTS frame, and simultaneously sends a busy tone in the BTt channel until the completion of the RTS to prevent interferers. Upon reception of the RTS, the data receiver sends a busy tone in the BTr channel, which serves the same function as CTS. Upon reception of the BTr busy tone, the data transmitter transmits its data-DATA frame. When the data receiver is receiving the data-DATA frame, it keeps sending a BTr busy tone to prevent interferers. If data-DATA frame is received successfully, the data receiver continues to send a BTr busy tone for a small duration (i.e., the busy-tone detection time), which serves as an ACK.

If the traffic source node does not receive the BTr busy tone after its transmission of an RTS or DATA frame, a collision is inferred. The source node will double its contention window (until the maximum contention window CW_{max} is reached), select a new backoff timer, and start its next contention after the two busy-tone channels have been sensed to be idle for its AIFS again. The contention window is reset to the initial value CW_{min} upon a successful transmission. Note that voice and data nodes keep the same CW_{min} and CW_{max} in our scheme (to be further discussed in Sect. 5.2.4), unlike the IEEE 802.11e.

Details of the operation procedure of the proposed MAC scheme are presented in the following subsection.

5.2.1 Operation Procedure of the Proposed MAC Scheme

Figure 5.1 illustrates the state transition diagram of the proposed MAC scheme. The ellipses represent the states of one node, and the name of each state transition is labeled along the path. At the initialization of the network, every node is at the *Idle* state. Detailed state transition procedure is as follows.

---

[2]In this book, a voice-DATA frame means a DATA frame from a voice traffic source, while a data-DATA frame means a DATA frame from a data traffic source.

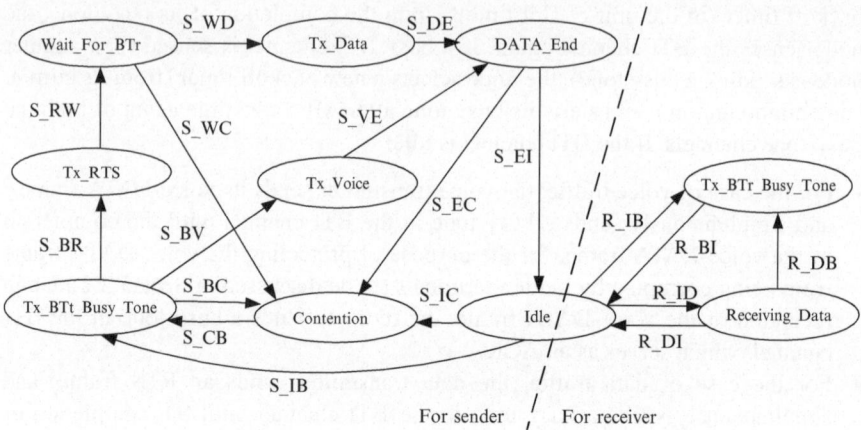

Fig. 5.1 The state transition diagram of the proposed dual busy-tone scheme

- *Transition S_IB/S_IC:* When a node is at the *Idle* state and has traffic to send, it sets its *CW* and AIFS according to the traffic type, and chooses a random backoff timer from [0, *CW*]. Then the node senses the BTr and BTt channels for the duration of AIFS[AC_voice] (or AIFS[AC_data]). If no busy-tone signal for AIFS[AC_voice] (or AIFS[AC_data]), the node will send a busy tone to jam the BTt channel, and the length of the busy tone (in the unit of slot time) is equal to its backoff timer. Then the node goes into the *Tx_BTt_Busy_Tone* state. If the node senses a busy-tone signal in either channel, it goes into the *Contention* state.
- *Transition S_BR/S_BV/S_BC:* When a node is at the *Tx_BTt_Busy_Tone* state, after the completion of its own BTt busy tone, the node monitors both BTt and BTr channels for one slot time. If both channels are idle (which means the node is sending the longest BTt busy tone), the node will transmit. If the node has a data packet to send, it transmits its RTS frame in the information channel, at the same time sends busy tone in the BTt channel, and goes into the *Tx_RTS* state; if the node has a voice packet to send, it transmits voice-DATA frame in the information channel, at the same time sends busy tone in the BTt channel, and goes into the *Tx_Voice* state. If either the BTt channel or the BTr channel is busy, the node goes into the *Contention* state.
- *Transition S_RW:* When a node is at state *Tx_RTS*, it keeps transmitting an RTS frame at the information channel and the BTt busy tone at the BTt channel. At the end of RTS frame transmission, the node stops its BTt busy tone, and sets a timer (equal to the busy-tone detection time), and goes into the *Wait_For_BTr* state.
- *Transition S_VE:* When a node is at state *Tx_Voice*, it keeps transmitting voice-DATA frame in the information channel and the BTt busy tone in the BTt channel. At the end of voice-DATA frame transmission, the node stops its BTt busy tone, and sets a timer (equal to the busy-tone detection time), and goes into the *DATA_End* state.
- *Transition R_ID/R_IB:* When a node is in the *Idle* state and has no backlogged traffic to send, it keeps monitoring the information channel to check if there is

any RTS (or voice-DATA) frame destined to it. When the node receives an RTS (or voice-DATA) frame, it sends a busy tone immediately in the BTr channel as an indication of successful reception. If the received frame is an RTS frame, it sets a timer (equal to the busy-tone detection time plus data-DATA frame transmission time), and goes into the *Receiving_Data* state. If the received frame is a voice-DATA frame, it goes into the *Tx_BTr_Busy_Tone* state.

- *Transition R_BI:* When a node is at state *Tx_BTr_Busy_Tone*, it continues its BTr busy tone for the busy-tone detection time, then stops the BTr busy tone and goes into the *Idle* state.

- *Transition S_WD:* When a node is at the *Wait_For_BTr* state, it senses the BTr channel. If a BTr busy tone is sensed, it sends a data-DATA frame immediately in the information channel, and goes into the *Tx_Data* state.

- *Transition S_WC:* When a node is at the *Wait_For_BTr* state, it senses the BTr channel. If the node does not sense a BTr busy tone (which means a collision may happen), upon timeout, the node doubles its *CW* (up to CW_{max}) and goes into the *Contention* state.

- *Transition S_DE:* When a node is at the *Tx_Data* state, it keeps transmitting its data-DATA frame. When the data-DATA transmission is finished, the node sets a timer (equal to busy-tone detection time), and goes into the *DATA_End* state.

- *Transition R_DB:* When a node is at the *Receiving_Data* state, it receives a data-DATA frame. If the data-DATA frame is successfully received, it goes into the *Tx_BTr_Busy_Tone* state.

- *Transition R_DI:* When a node is at state *Receiving_Data*, it receives a data-DATA frame. If it does not successfully receive a data-DATA frame, upon timeout, it stops the BTr busy tone immediately and goes into the *Idle* state.

- *Transition S_EI/S_EC:* When a node is at the *DATA_End* state, it senses the BTr channel. If it senses a BTr busy tone (which means the destination successfully receives the data frame), the node resets its *CW* to the initial value CW_{min}, and goes into the *Idle* state. If the node does not sense a BTr busy tone in the *DATA_End* state, upon timeout, it doubles its *CW* (up to CW_{max}) and goes into the *Contention* state.

- *Transition S_CB:* When a node is at the *Contention* state, it randomly chooses a backoff timer from its current *CW*. Then the node senses the BTr and BTt channels for the duration of AIFS[AC_voice] (AIFS[AC_data]). If no busy tone is sensed, the node will send a busy tone to jam the BTt channel, and the length of the busy tone (in the unit of slot time) is equal to its backoff timer. Then the node goes into the *Tx_BTt_Busy_Tone* state. If the node senses a busy-tone signal in either channel, it remains at the *Contention* state.

5.2.2 Solution to the Hidden Terminal Problem

To achieve good performance, not only data-DATA frame collisions but also RTS and voice-DATA frame collisions caused by hidden terminals should be completely avoided if possible. In our MAC scheme, the use of an increased carrier sense range in the BTt channel can help to achieve this target. To protect an RTS (or a voice-DATA) frame from being corrupted by hidden terminals, when a sender starts to transmit its frame, it also transmits a busy tone in the BTt channel, and stops it when the RTS (or voice-DATA) frame transmission is finished. Because of the increased carrier sense range of the BTt channel, all the potential hidden terminals that may interfere with this ongoing transmission can sense the BTt channel being busy, and thus defer their own transmissions and avoid corrupting the RTS (or voice-DATA) frame transmission. Further, for data traffic, when the sender completes the RTS transmission and the destination node recognizes that it is the intended receiver, the destination will send a busy tone immediately in the BTr channel (i.e., serves the same function as CTS). All the potential hidden terminals of the sender can hear this busy tone, thus deferring their transmissions. The destination continues sending the BTr busy tone during the whole data-DATA frame reception. Therefore, collision is avoided from the beginning of the RTS transmission to the end of the data-DATA frame transmission.

5.2.3   Solution to the Exposed Terminal Problem

With the use of the BTr busy-tone channel, our scheme can resolve the exposed terminal problem. When a desired receiver receives an RTS (or DATA) frame, instead of responding with a CTS (or ACK) frame in the information channel, the receiver sends a busy tone in the BTr channel that serves the same function as an CTS (or ACK) frame. After sending out an RTS (or DATA) frame, the sender senses the BTr channel. The status of a busy BTr channel indicates that the RTS (or DATA) frame has been successfully received by the receiver; otherwise, a collision has occurred. Replacing the CTS and ACK frames with the BTr busy tones allows multiple senders within one-hop neighborhood to send their frames simultaneously (as long as they do not interfere with each other at the receivers) without the problem that the feedback from the receiver may be corrupted by other ongoing DATA transmissions, since the feedback and DATA transmissions are in different channels.

To replace the CTS and ACK frames with the BTr busy tones, it is essential to ensure that, when a sender senses a BTr busy tone after completing its frame transmission, this busy tone must be from its own destination rather than from any other nodes (since the BTr busy tone does not carry any information). This is achieved in our scheme as follows. When a sender is sending an RTS (or DATA) frame, all the potential receivers (which are the destinations of other nodes) within this sender's one-hop neighborhood cannot correctly receive their own frames; and therefore, none of them will send the BTr busy tone as a feedback.

5.2.4   Solution to the Priority Reversal Problem

To address the priority reversal problem, it is desired to ensure the channel access priority for voice traffic independent of the node location. Our approach is to let all potential hidden terminals of the voice node be aware that the voice node is contending for the channel, so that they defer their own contentions. This is achieved with the use of the BTt busy tone. In our scheme, after waiting for both the busy-tone channels to be idle for an AIFS[voice], a voice node sends a BTt busy tone. Thus, for data nodes, if there exists one or more voice contenders within its two-hop neighborhood, they will sense the BTt busy tone (from voice nodes) during the AIFS[data] (>AIFS[voice]), and defer their transmissions. Therefore, the voice node avoids the priority reversal problem no matter where it is located, benefiting from the doubled carrier sensing range of the BTt channel.

Note that the proposed MAC scheme not only avoids priority reversal in a non-fully-connected environment, but also ensures guaranteed priority access for voice traffic. Although IEEE 802.11e and our scheme use the same AIFS settings, our scheme can achieve guaranteed priority access for voice over data in each contention, while IEEE 802.11e can only achieve statistic priority access for voice over a long term. The advantage of our scheme comes from the different BTt busy tone starting moments of voice and data nodes. This also explains why the same CW_{min} and CW_{max} are adopted by voice and data nodes in our scheme, unlike the IEEE 802.11e.

5.2.5 Solution to the Unfairness Problem

In 802.11e, the node with the smallest backoff timer transmits. When a node transmits successfully, its contention window is reset to the initial value, and thus its chance to win the next contention is still large. When a packet transmission is collided, the contention window of the source node is doubled (up to the maximum value), and thus its chance to win the next contention is small.

On the contrary, in our scheme, the nodes with the largest backoff timer transmit. Upon a successful transmission, the node's contention window is reset to the initial value, so that its chance to have the largest backoff timer among all the nodes and win the next contention is small. Upon a collision of its packet, the node's contention window is doubled (up to the maximum value), so it has a large chance to have the largest backoff timer among all the nodes and win the next contention. This means that our scheme can distribute the channel access time more fairly to the contending nodes than 802.11e. See the example in Fig. 2.3c in Chap. 2. Suppose that after one contention, node C resets its CW to CW_{min} (because of a successful transmission) while node A doubles its CW (because of the failed RTS), then in the next contention, it is very likely that node A has a larger backoff timer than node C, thus sends a longer BTt busy tone (which can be heard by node C). Therefore, node A will win the access to the channel and node C will defer its transmission.

It may seem that the waiting time (before getting the channel) of a node is longer with our scheme than that with IEEE 802.11e, since the node with the largest backoff timer instead of the smallest backoff timer (as in IEEE 802.11e) gets the channel. However, as shown in our another work [42], the CW setting can be set to small values in our scheme. The CW setting {3:15} (i.e., $CW_{min} = 3$ and $CW_{max} = 15$) works well for a large number (up to 500) of contending nodes, and the throughput and collision probability are quite stable when the number of contending nodes increases. Therefore, the negative effect of the longer backoff time can be neglected. The insensitivity of the system performance to the number of nodes also facilitates network configuration.

The idea to let the node with the longest busy tone wins the channel is inspired by the black-burst scheme [75]. However, the original idea of black-burst is proposed to provide QoS guarantee for real-time traffic, and cannot be directly applied to solve the unfairness problem. Here we adopt the "jamming" nature of black-burst, and modify the backoff procedure (as discussed above). A result of the modification is good fairness performance. Furthermore, in the black-burst scheme, only two traffic classes are supported. In our scheme, different classes are differentiated by different AIFSs, and the black burst length is determined by the backoff timer value. So although we only consider two traffic classes (i.e., voice and data), our scheme can support more traffic classes as long as they have different AIFS values.

Till now, the proposed scheme is considered in a stationary network environment. It can also work well in a mobile network. Since the BTr busy tone is sent by a receiver during the whole data packet reception, and can be heard by all the potential interferers no matter whether they move or not, any potential sender which is in the vicinity of an existing receiver will defer its own transmission to avoid collision even in a mobile environment. A receiver's data packet reception may be corrupted only in the scenario where a node has started its transmission when it is far away from the receiver, and before it finishes its transmission, it moves to the vicinity of the receiver. However, this scenario rarely happens since a data packet transmission time is too short to allow a node to move a long distance. For example, it takes about 4 ms to transmit a 1,000-byte data packet at a 2 Mpbs channel rate. For a node with velocity of 60 km/h, it will move about 0.07 m during 4 ms. Considering an RTS frame, it has an even smaller size (i.e., 20 bytes), so the possibility that RTS collisions happen due to mobility can be negligible. From the above discussion, we can see that the effectiveness of the proposed scheme will not be affected by the mobility of the nodes as long as their velocities are not extremely high. To deal with the extreme case when the velocity is very high, we can let each sender monitor the BTr channel during its transmission. When a busy BTr channel is detected (i.e., a nearby node is receiving a packet), the sender immediately stops its transmission to avoid collision.

5.3   Performance Analysis

In this section, we present the analysis of data throughput in our proposed scheme in a single-hop case without voice traffic. As the analytical result for a multi-hop network with voice traffic is difficult to obtain, we resort to extensive simulations for the performance in a multi-hop network.

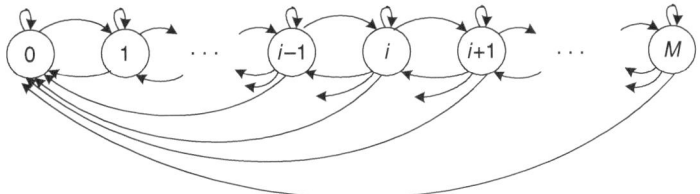

Fig. 5.2 The state transition diagram of $m(t)$

Consider M data source nodes. For simplicity of presentation, the contention window of each data node takes values from the set $\{CW_1^d, CW_2^d\}$ (i.e., $CW_{\min}[\text{data}] = CW_1^d$, $CW_{\max}[\text{data}] = CW_2^d$) where $CW_2^d = 2 \cdot (CW_1^d + 1) - 1$. Our analysis can be easily extended to the cases with more choices of contention windows. Let $m(t)$ denote the number of data nodes with contention windows size CW_1^d at time instant t, and therefore $M - m(t)$ data nodes are with contention windows size CW_2^d. Define a *transmission event* as a successful transmission or a collision. We sample the value of $m(t)$ at the beginning of each transmission event, and form a discrete-time Markov process, as shown in Fig. 5.2.

For state $m(t) = i$, let j_1 and j_2 denote the numbers of nodes that transmit in the next transmission event with contention windows CW_1^d and CW_2^d, respectively.[3] The probability of the largest backoff timer value l in such a transmission event (i.e., j_1 nodes with CW_1^d and j_2 nodes with CW_2^d choose a backoff timer l and all other nodes choose backoff timers less than l) is given by

$$
p_{j_1, j_2, l|i} = \begin{cases} \binom{i}{j_1}\left(\frac{1}{CW_1^d+1}\right)^{j_1}\left(\frac{l}{CW_1^d+1}\right)^{i-j_1} \cdot \binom{M-i}{j_2}\left(\frac{1}{CW_2^d+1}\right)^{j_2}\left(\frac{l}{CW_2^d+1}\right)^{M-i-j_2}, \; 0 \le l \le CW_1^d, \\ \qquad\qquad\qquad \text{if } j_1 \ne 0 \\ \min\{1, \left(\frac{l}{CW_1^d+1}\right)^i\} \cdot \binom{M-i}{j_2}\left(\frac{1}{CW_2^d+1}\right)^{j_2}\left(\frac{l}{CW_2^d+1}\right)^{M-i-j_2}, \; 0 \le l \le CW_2^d, \\ \qquad\qquad\qquad \text{if } j_1 = 0 \end{cases}
$$

$$(5.1)$$

where $j_1 + j_2 \ge 1$. The event is a successful transmission if $j_1 + j_2 = 1$, or collision if $j_1 + j_2 > 1$.

Figure 5.2 illustrates the state transition diagram of $m(t)$. For state $m(t) = i$, after a transmission event, the process will

- Remain at state i if one node with CW_1^d transmits successfully (i.e., $j_1 = 1$ and $j_2 = 0$) with probability $\sum_{0 \le l \le CW_1^d} p_{1,0,l|i}$, or a collision happens in which no node with CW_1^d but at least two nodes with CW_2^d are involved (i.e., $j_1 = 0$ and $j_2 \ge 2$) with probability $\sum_{0 \le l \le CW_2^d, \, 2 \le j_2 \le M-i} p_{0,j_2,l|i}$;
- Transit to state $i+1$ if one node with CW_2^d transmits successfully (i.e., $j_1 = 0$ and $j_2 = 1$), with probability $\sum_{0 \le l \le CW_2^d} p_{0,1,l|i}$;

---

[3] Here we omit the time index t for j_1 and j_2.

- Transit to state $i - k$ $(1 \leq k \leq i)$ if a collision happens in which k nodes with CW_1^d are involved, with probability $\sum_{0 \leq l \leq CW_1^d, \, 1 \leq j_2 \leq M-i} p_{1,j_2,l|i}$ when $k = 1$, or $\sum_{0 \leq l \leq CW_1^d, \, 0 \leq j_2 \leq M-i} p_{k,j_2,l|i}$ when $k > 1$.

Based on the transition probabilities among the states in Fig. 5.2, we can obtain the steady-state probabilities of all the states, $[\pi(0), \pi(1), \ldots, \pi(M)]$. Let t_s and t_c denote the times of a successful transmission and a collision (not including the backoff time), respectively, given by

$$\begin{cases} t_s = \text{AIFS[AC\_data]} + S_{\text{RTS}}/R_{\text{basic}} + t_{\text{det}} + S_{\text{d\_DATA}}/R + t_{\text{det}} \\ t_c = \text{AIFS[AC\_data]} + S_{\text{RTS}}/R_{\text{basic}} + t_{\text{det}} \end{cases} \quad (5.2)$$

where S_{RTS} and $S_{\text{d\_DATA}}$ are the RTS and data-DATA frame sizes in bits, respectively, R_{basic} and R are the basic rate (for RTS transmission) and information transmission rate (for DATA transmission), and t_{det} is BTr busy-tone detection time (i.e., the sender detects the BTr busy tone after an RTS and/or a data-DATA frame transmission). Then the average time in a transmission event of state i is

$$\overline{t_e(i)} = \sum_{l; \, j_1+j_2=1} p_{j_1,j_2,l|i} \cdot (l \cdot \tau + t_s) + \sum_{l; \, j_1+j_2>1} p_{j_1,j_2,l|i} \cdot (l \cdot \tau + t_c). \quad (5.3)$$

where τ is the slot time. Thus, we can calculate the average system throughput as
$$T_s^a = \frac{S_{\text{d\_DATA}} \cdot \sum_{0 \leq i \leq M} \pi(i) \cdot \sum_{l; \, j_1+j_2=1} p_{j_1,j_2,l|i}}{\sum_{0 \leq i \leq M} \pi(i) \cdot t_e(i)}.$$

5.4 Performance Evaluation

To evaluate the performance of our proposed scheme, we compare it with IEEE 802.11e and DBTMA. We compare our scheme with IEEE 802.11e in all cases and with DBTMA in the cases with hidden and exposed terminals, since DBTMA focuses on the issues of hidden and exposed terminal problems, but not on priority and fairness issues. Since DBTMA does not explicitly specify its backoff mechanism, for fair comparison, we use the same backoff mechanism for DBTMA as that in our scheme. For DBTMA, only data traffic is considered. We choose the GSM 6.10 codec as the voice source as an example. Long-lived data traffic is considered (each data node always has frames to send).

As shown in [42], both voice and data traffic can choose the CW setting $\{3{:}15\}$ in our scheme. For IEEE 802.11e and DBTMA, it is not appropriate to use the same small CW sizes as those in our scheme. For 802.11e, a small CW setting leads to serious collisions and low throughput when the number of contending nodes increases. For DBTMA, it leads to a very low throughput in the network with hidden terminals. The simulation parameter values are listed in Table 5.1. First, the system performance is evaluated under some specific network topologies as shown in Fig. 5.3. Then random topologies are simulated for more comprehensive evaluation.

Parameter	Value
Slot time τ	20 μs
SIFS	10 μs
AIFS[AC_voice]	30 μs
AIFS[AC_data]	50 μs
CW_{min} (for the proposed scheme)	3
CW_{max} (for the proposed scheme)	15
CW_{min}[voice] (for IEEE 802.11e)	15
CW_{max}[voice] (for IEEE 802.11e)	127
CW_{min}[data] (for IEEE 802.11e)	31
CW_{max}[data] (for IEEE 802.11e)	1,023
CW_{min} (for DBTMA)	15
CW_{max} (for DBTMA)	255
PHY preamble	192 μs
MAC header	36 bytes
S_{RTS} (RTS frame size)	20 bytes
S_{CTS} (CTS frame size)	14 bytes
S_{ACK} (ACK frame size)	14 bytes
S_{d_DATA} (data-DATA frame size)	1,000 bytes
Link rate (for IEEE 802.11e)	11 Mbps
Information channel rate (for the proposed scheme)	10.9 Mbps
Basic rate (for RTS and/or CTS transmission)	2 Mbps
t_{det} (busy-tone detection time)	10 μs

Table 5.1 Simulation parameters used for the dual busy-tone MAC scheme

5.4.1 Throughput in a Scenario with Hidden Terminals

We compare the performance in scenarios (a) and (b) (see Fig. 5.3). Scenario (a) is a fully-connected network with $N_a (= 4, 12, 20, \ldots)$ senders. Scenario (b) is a network with hidden terminals, which has four groups,[4] each group containing $N_a/4$ senders. In both scenarios, all the senders send data traffic to a common receiver. Figure 5.4 shows the RTS collision probability (which is approximated by the ratio of the collided RTS frame number to the total transmitted RTS frame number). For IEEE 802.11e and DBTMA, the RTS collision probability in scenario (b) is much higher than that in scenario (a). Correspondingly, the aggregate throughput in scenario (b) is much lower than that in scenario (a), shown in Fig. 5.5. The gap is contributed by the hidden terminals. On the contrary, with our scheme, the RTS collision probability and the aggregate throughput almost remain the same in both scenarios. The hidden terminals in scenario (b) do not introduce more RTS collisions, indicating that our scheme effectively avoids RTS collisions caused by

---

[4]Each group represents a set of nodes, which are in the transmission range of each other and are contending with each other. The nodes in the same group have the same characteristics. The nodes in one group are beyond the transmission range of any node in other groups.

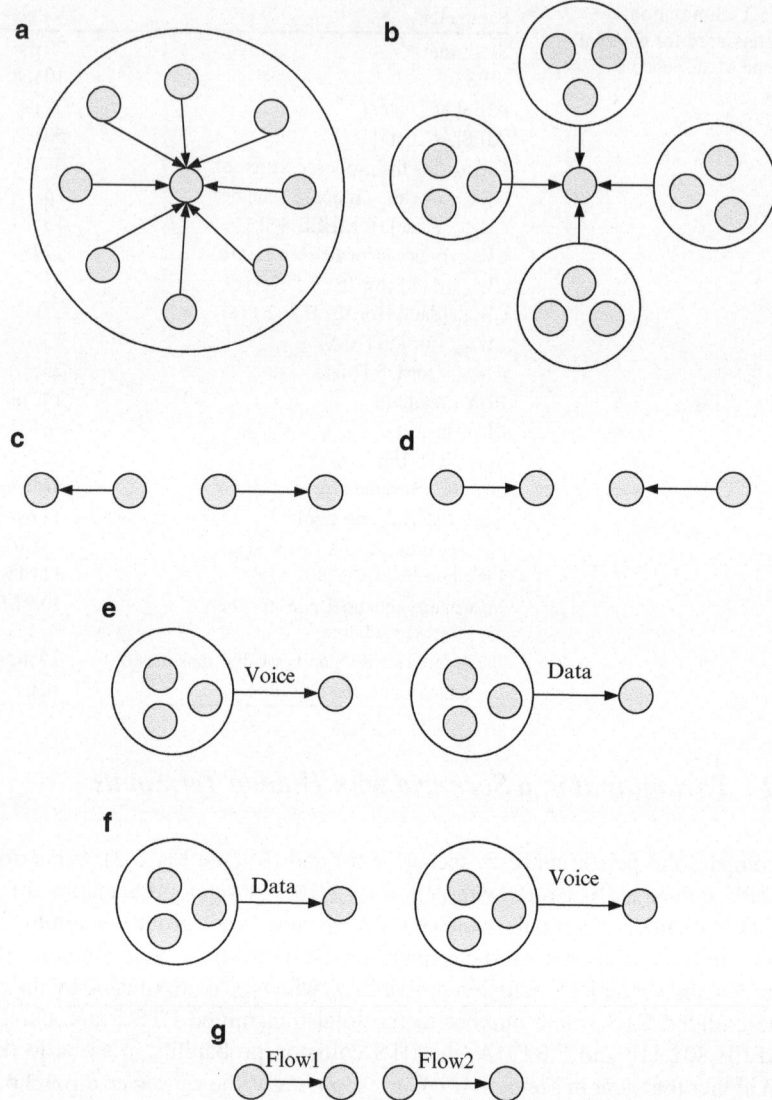

Fig. 5.3 The network topologies used in simulation of the wireless ad hoc network

hidden terminals. Note that completely avoiding RTS collisions caused by hidden terminals does not mean that the RTS collisions do not happen. Actually, in scenario (a) without hidden terminals, RTS collisions still exist. Such collisions happen when more than one contending nodes choose the same backoff timer. We resolve those collisions by doubling the *CW* of the collided nodes as in IEEE 802.11e. A large *CW* results in a small RTS collision probability but a long backoff time. In order to

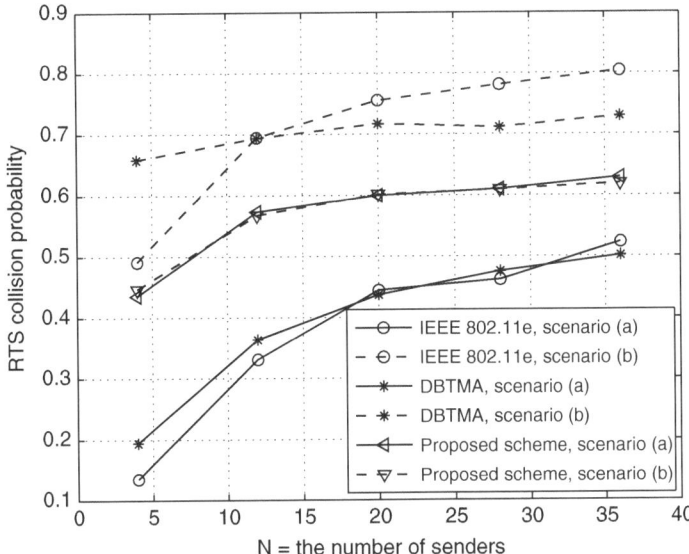

Fig. 5.4 RTS collision probability for IEEE 802.11e, DBTMA, and our scheme in scenarios (a) and (b)

maximize the resource utilization, there exists a tradeoff between the RTS collision probability and the backoff time. Hence, it is not necessary to eliminate such RTS collisions. However, it is desirable to reduce the RTS collisions caused by hidden terminals to zero if possible, since they reduce the efficiency of resource utilization.

From Fig. 5.4 we notice that, in scenario (a), the RTS collision probability of our scheme is higher than that of IEEE 802.11e and DBTMA. It is because our scheme uses a smaller contention window size than those for IEEE 802.11e and DBTMA (see Table 5.1). The smaller the contention window, the higher the RTS collision probability. However, the backoff time is also reduced significantly in our scheme. As a result, the aggregate throughput in our scheme is still higher than those of IEEE 802.11e and DBTMA, as shown in Fig. 5.5.

5.4.2   Throughput in Scenarios with Exposed Terminals

In scenario (c) in Fig. 5.3 the senders are exposed terminals, while in scenario (d) the receivers are exposed terminals. In both scenarios, senders send data traffic to the corresponding receivers. Table 5.2 compares the aggregate throughput of the proposed scheme, IEEE 802.11e and DBTMA in these two scenarios. For comparison, the throughput of a single data flow (i.e., only a data flow exists in the network) is also presented. We can see that the aggregate throughput of IEEE 802.11e in scenarios (c) and (d) is similar to the single flow throughput, indicating

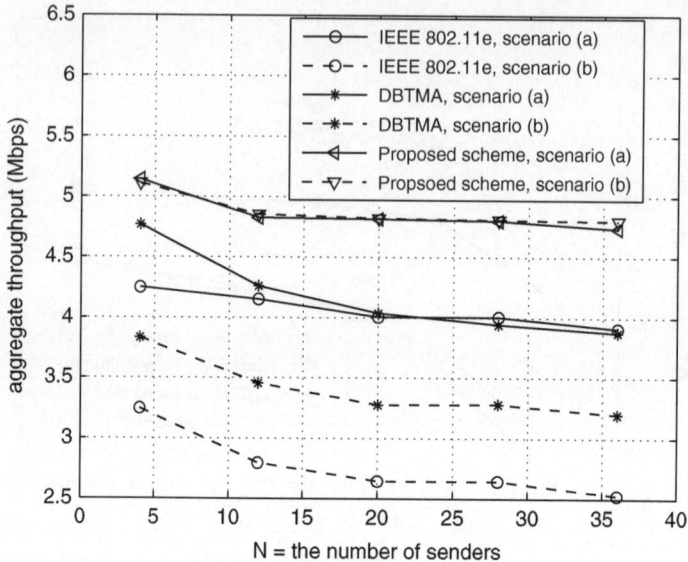

Fig. 5.5 The aggregate throughput of IEEE 802.11e, DBTMA, and our scheme in scenarios (a) and (b)

Table 5.2 The aggregate throughput (Mbps) in scenarios (c) and (d) shown in Fig. 5.3	Scenario (c)	Scenario (d)	Single-flow
Proposed	11.95	11.96	5.98
IEEE 802.11e	4.22	3.84	3.87
DBTMA	10.87	10.97	5.48

that IEEE 802.11e suffers from the exposed terminal problem. On the contrary, the aggregate throughput of our scheme and DBTMA in scenarios (c) and (d) are almost two times the single-flow throughput, indicating that our scheme and DBTMA allow simultaneous transmissions among exposed terminals.

5.4.3   Priority Access

First, we evaluate the priority performance of the proposed scheme supporting voice/data traffic in a fully-connected network. As voice traffic is delay-sensitive, frames with a large delay are considered useless and discarded. In the simulation, we set the voice frame delay bound as 40 ms. If a voice frame cannot be delivered successfully within the delay bound after its generation, it will be dropped by the voice sender. Figure 5.6 shows the voice frame dropping probability in different MAC schemes, for 20 voice source nodes when the number of data source nodes changes from 10 to 60. No voice frame dropping is observed in our scheme, while

Fig. 5.6 Voice frame
dropping probability versus
data node number in a
fully-connected network with
20 voice nodes

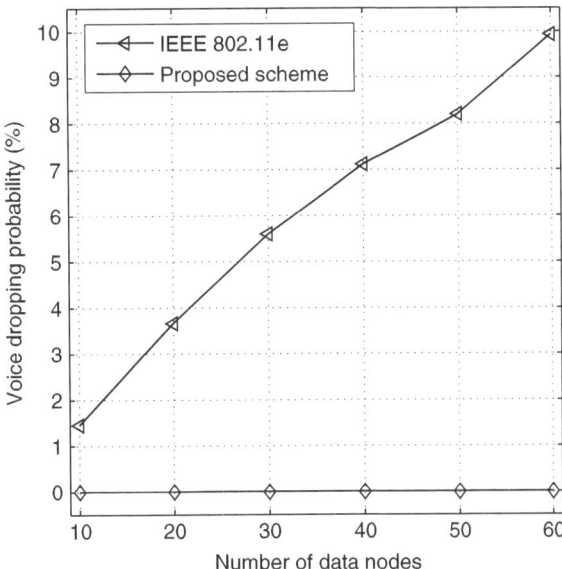

Table 5.3 The average voice access delay (ms) with different node number N_g within a group in scenarios (e) and (f) shown in Fig. 5.3

N_g		1	2	3	4	5	6	7	8	10	20
802.11e	Scenario (e)	9.9	27.1	45.7	57.3	97.1	101.6	165.3	175.5	228.3	686.0
	Scenario (f)	1.38	1.17	1.27	1.19	1.31	1.21	1.26	1.24	1.25	1.32
Proposed	Scenario (e)	1.27	1.15	1.11	1.11	1.12	1.10	1.08	1.11	1.11	1.11
	Scenario (f)	1.09	1.07	1.06	1.16	1.06	1.10	1.10	1.15	1.11	1.13

in IEEE 802.11e the voice frame dropping probability increases with the data node number. The results indicate that our proposed scheme (which provides guaranteed priority access) has better QoS provisioning capability than IEEE 802.11e (which provides statistical priority access).

Next, we choose two specific scenarios (e) and (f) (see Fig. 5.3) to study whether or not the priority access is dependent on the locations of the flows. In both scenarios, a group of voice nodes are contending with a group of data nodes. Each group contains N_g nodes, sending traffic to a common receiver. We use the average voice flow access delay (which is the time duration from the moment that the frame is at the top of the buffer to the moment that the frame has been successfully transmitted) as the performance metric, given in Table 5.3. For IEEE 802.11e, the voice access delay is quite large in scenario (e), from 9.9 to 863.7 ms with the increase of N_g; while in scenario (f), the delays are around 1 ms for all the N_g values. These results indicate that the priority access performance of IEEE 802.11e is location-dependent. On the contrary, in our scheme, the voice access delay almost remains the same (around 1 ms) in both scenarios and for all the N_g values, indicating that our scheme provides a stable priority access, independent of the flow locations.

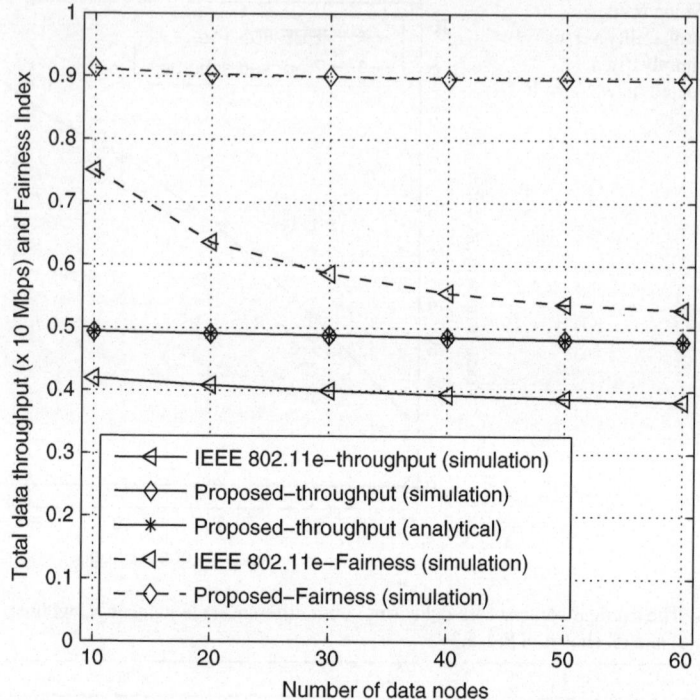

Fig. 5.7 The total data throughput and Fairness Index versus data node number in a fully-connected network with only data traffic

5.4.4 Fairness

First, we compare the short-term fairness performance of IEEE 802.11e and our scheme in a fully-connected network as shown in scenario (a) in Fig. 5.3. The fairness is measured by Jain's Fairness Index given by $\frac{(\sum_{i=1}^{K_d} T_i)^2}{K_d \cdot \sum_{i=1}^{K_d} T_i^2}$ [40], where T_i is the throughput of the ith data node over a time window, and K_d is the number of data nodes. The higher the Fairness Index value, the better the fairness performance. We sample the Fairness Index values after each duration over which each data node transmits six frames on average. Figure 5.7 compares the average Fairness Index values. As expected, our scheme shows better short-term fairness performance than IEEE 802.11e. The aggregate throughputs of the two schemes are also shown in Fig. 5.7. We can see that the analytical and simulation results for our scheme match well. The aggregate throughput of our scheme is larger than that of IEEE 802.11e.

Next, we compare the long-term fairness performance of IEEE 802.11e and our scheme in scenario (g) in Fig. 5.3 (since even long-term fairness is difficult to achieve in this scenario, we do not consider short-term fairness, but compare the achieved throughput of each flow). It is found that flows 1 and 2 achieve a throughput of 3.46 and 2.31 Mbps, respectively, in our scheme, but 0.2 and

3.77 Mbps, respectively, in IEEE 802.11e. In IEEE 802.11e, flow 1 is almost starved while flow 2 occupies the channel almost all the time; in our scheme, each flow gets a certain share of the channel time. Note that our scheme improves the long-term fairness performance to some degree as compared with IEEE 802.11e, but not yet achieves absolute fairness. To achieve absolute fairness in a distributed manner is extremely challenging. Extra information needs to be exchanged among the nodes, and a controller is needed to coordinate the transmissions from the nodes, making the scheme not scalable. Therefore, there exists a tradeoff between scalability and absolute fairness.

5.4.5 *Performance in Random Topologies*

We consider a $1,000 \times 1,000\,m^2$ service area, where the transmission range of each node is 200 m. The nodes are evenly distributed in the whole area. The flows are randomly chosen from the nodes which are one hop away. Half of the flows are voice flows and the remaining are data flows. Here we simulate three cases: sparse (36 nodes with 10 flows), medium (121 nodes with 50 flows), and dense (441 nodes with 200 flows). The node density is measured as d, where the number of nodes $N_n = d^2$. Initially, we choose $d = 6$ for sparse case. We increases d by 5 for medium case, and further by 10 for dense case. In our experiment, we use the different node densities to reflect the different contention degrees of the network. When the node density increases within a fixed area (i.e., the number of nodes increases), the number of flows (and traffic load) will increase, so does the contention degree.

To compare the priority access performance, we show the CDF (cumulative distribution function) of voice access delay in Fig. 5.8. The vertical axis is the probability that the voice access delay is larger than the delay specified in the horizontal axis. It is clear that our scheme has a smaller voice access delay than IEEE 802.11e in all the three cases. In the sparse case, all voice frames' access delays are below 5 ms in our scheme, and around 11% voice frames have an access delay larger than 5 ms in IEEE 802.11e. With the increase of user density, the voice access delay increases in both schemes because the voice flows encounter a much higher contention level and are more likely to collide. In the dense case, around 30% (10%) voice frames have an access delay larger than 10 ms in IEEE 802.11e (our scheme). The aggregate data traffic throughput of IEEE 802.11e is 14.56, 21.77, and 11.54 Mbps in the sparse, medium, and dense cases, respectively, while for our scheme it is 23.41, 46.16, and 30.21 Mbps, respectively. That is, our scheme has a higher throughput than IEEE 802.11e in all the three cases. Note that the system throughput in the medium case is larger than those in the other two cases. In the sparse case, as a small number of flows contend for the channel, the network capacity is not fully utilized. With an increased flow number in the medium case, the system throughput increases. When the flow number further increases (in the dense case), more resources are used by voice traffic, resulting in a reduced throughput of data traffic.

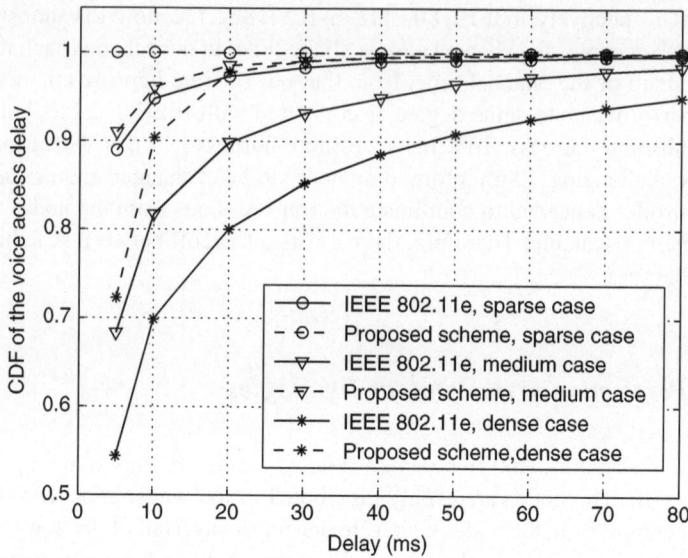

Fig. 5.8 The CDF of voice access delay in random topologies

5.4.6   Sensitivity of the Proposed Scheme to Carrier Sense Ranges

With an appropriate carrier sense range setting, the hidden/exposed terminal problem, priority reversal and unfairness problems are eliminated and, at the same time, the resources are efficiently utilized. In reality, the carrier sense ranges may not be set exactly as required, resulting in reduced efficiency or effectiveness of the proposed scheme. To investigate the sensitivity of the proposed scheme to different carrier sense ranges, random topologies with different node densities are considered. The random topologies and flows are generated in the same way as in the preceding subsection. BTt channel's carrier sense range is set to θ_1 times transmission range, where θ_1 varies from 1.6 to 2.4. BTr channel's carrier sense range[5] is set to θ_2 times transmission range, where θ_2 varies from 1.0 to 1.2.

Table 5.4 compares the average voice packet delays and the aggregate data throughputs with different BTt/BTr carrier sense range settings in the networks with different node densities. It can be seen that the average voice access delay changes slightly with different carrier sense range settings. For the aggregate data throughput, it remains almost the same in the sparse case, and reduces slightly in the medium and dense cases when the BTt/BTr channel's carrier sense range increases. In the medium and dense cases, when the BTt channel's carrier sense range is

---

[5]Typically the carries sense range is no less than the transmission range, so we do not consider the case that BTr channel's carrier sense range is less than the transmission range.

Table 5.4 The average voice access delay (ms) and aggregate data throughput (Mbps) with different carrier range settings

θ_1 and θ_2 values		θ_1=1.6 θ_2=1.0	θ_1=1.8 θ_2=1.0	θ_1=2.0 θ_2=1.0	θ_1=2.2 θ_2=1.0	θ_1=2.4 θ_2=1.0	θ_1=2.0 θ_2=1.1	θ_1=2.0 θ_2=1.2
Sparse case	Voice access delay	0.66	0.67	0.74	0.74	0.80	0.74	0.75
	Data throughput	23.57	23.56	23.41	23.41	22.80	23.41	23.42
Medium case	Voice access delay	2.75	2.64	2.50	2.50	2.57	2.68	2.29
	Data throughput	49.38	49.47	46.16	42.86	41.42	46.15	35.7
Dense case	Voice access delay	5.76	5.07	5.05	4.73	4.95	4.64	3.95
	Data throughput	35.74	33.15	30.21	26.07	21.27	28.94	27.72

larger than the coverage of two-hop neighborhood, some nodes may unnecessarily defer their transmissions, resulting in a reduced resource utilization. When the BTt channel's carrier sense range is less than the coverage of two-hop neighborhood, a slightly higher data throughput is achieved with the cost of unfairness. When the BTr channel's carrier sense range is larger than the coverage of one-hop neighborhood, the receivers may prevent some nodes (which may not corrupt their receptions) from transmitting concurrently. However, in the sparse case, since flows are likely to be far away from each other, the carrier sense ranges have little impact on the resource utilization.

5.5 Summary

In this chapter, we present a novel busy-tone based distributed MAC scheme supporting voice and data traffic in wireless ad hoc networks. Although many busy-tone based MAC scheme have been proposed in the literature, most of them are designed to solve hidden (or exposed) terminal problem. The newly proposed scheme is the first one to utilize the busy tones to address not only the hidden/exposed terminal problems, but also the priority reversal and unfairness problems associated with wireless ad hoc networks. The simulation results demonstrate that the system throughput is significantly increased by resolving the hidden and exposed terminal problems. As compared with the IEEE 802.11e, our scheme greatly reduces voice traffic delay by ensuring location-independent guaranteed priority access for voice traffic, and significantly improves fairness performance for data traffic.

Chapter 6
Collision-Free MAC for Wireless Mesh Backbones

In this chapter, we study a wireless mesh backbone, which consists of a number of routers located at fixed sites and covers a large geographical area. Different from the existing MAC schemes, our MAC scheme design benefits greatly from the fixed network topology. With the router location information, collision-free transmissions are scheduled in a deterministic way, without the request to RTS/CTS handshaking prior to every packet transmission. Thus, the overhead is greatly reduced, as compared with contention-based MAC schemes. Meanwhile, the deterministic schedule in our MAC scheme is adaptive to the traffic dynamic and can achieve maximal spatial frequency reuse. By eliminating collisions, reducing overhead, and achieving maximal spatial frequency reuse, the proposed scheme achieves much higher resource utilization than contention-based MAC schemes. Unlike most of the existing MAC schemes which are limited to single-hop communications, the proposed MAC scheme takes the end-to-end QoS provisioning for multi-hop flows into consideration.

6.1 Wireless Mesh Network

A typical example of the wireless mesh network consists of wireline gateways, wireless routers, and mobile stations, organized in a three-tier architecture [7, 44], as shown in Fig. 6.1. The third tier is the wireless access networks, through which users access the Internet. Wireless access networks includes WLANs, ad hoc networks, and cellular networks, among which the mobile users can seamlessly roam. The second tier is the wireless mesh backbone, consisting of a number of wireless routers at fixed sites. Each wireless router not only delivers traffic from the access networks in its coverage, but also forwards the traffic from and to its neighboring routers. The first tier is the mesh gateways, which connect the wireless mesh backbone to the Internet backbone. Normally a wireless mesh network covers a large geographical area. Thus, multi-hop communications are usually necessary,

P. Wang and W. Zhuang, *Distributed Medium Access Control in Wireless Networks*, SpringerBriefs in Computer Science, DOI 10.1007/978-1-4614-6602-4_6,

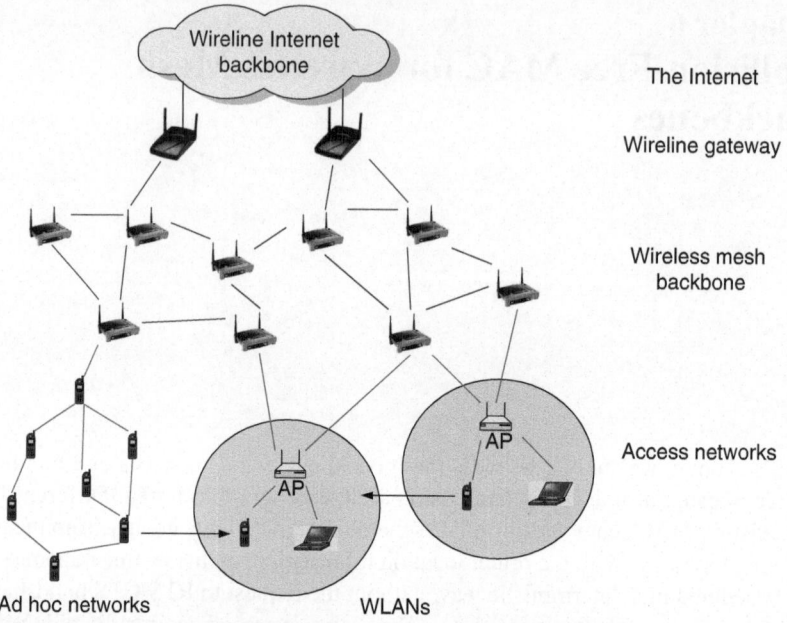

Fig. 6.1 An architecture of a broadband wireless mesh network

where a traffic flow from a source to its far away destination traverses multiple intermediate routers. Here we consider a single-channel wireless mesh backbone. The large scale of the backbone requires the MAC scheme to be scalable such that, when the network scale increases, the complexity and overhead of the MAC scheme do not increase dramatically, and the network performance does not degrade significantly.

6.2   The Distributed MAC Scheme

In our system, all routers are synchronized in time. There is a single information channel in the network, through which all the routers send their packets. Two routers are one-hop neighbors with each other if they are within the transmission range of each other. Based on the fixed locations of routers, the transmission power and rate for each wireless link can be appropriately determined, so that the required transmission accuracy at each link can be achieved and two or more links (which are more than two-hop away[1]) can transmit simultaneously without corrupting each other's transmissions.

---

[1]Two links are two-hop away when the receiver of each link is two-hop away from the source of the other link.

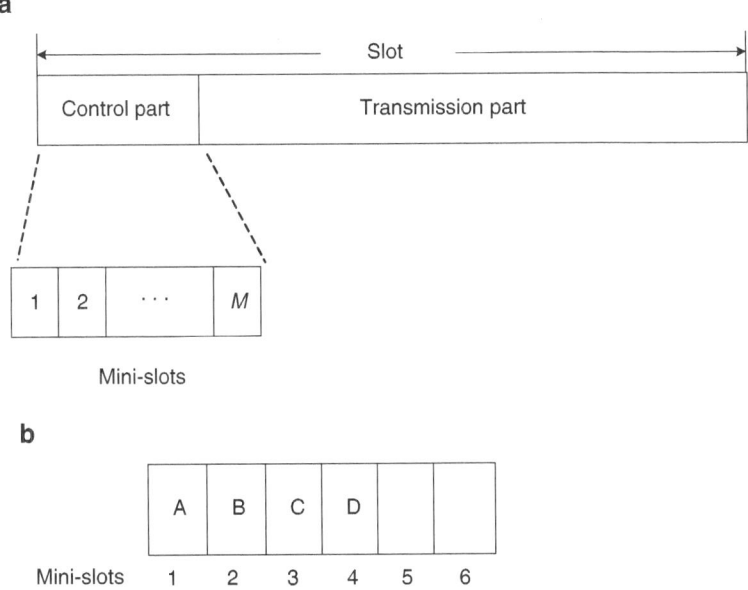

Fig. 6.2 The slot structure in the proposed collision-free MAC scheme

6.2.1 Distributed Time Slot Allocation

Time is partitioned into slots of constant duration, which are allocated to each router in a distributed manner. Once a router is allocated a slot, it can transmit one (or multiple) packet(s) to one (or multiple) one-hop neighbor(s), and all its one-hop and two-hop neighbors are not allocated the same slot in order to avoid packet transmission corruption. The same slot can be allocated to the routers which do not interfere with each other to achieve spatial frequency reuse. As shown in Fig. 6.2a, one slot consists of two portions: the first portion is the control part, occupying a very small fraction of the whole slot time. The control part is used to determine whether or not a router can transmit its packets in that slot; the second portion is the transmission part, dedicated to packet transmissions. The control part is further divided into several mini-slots, indexed sequentially with numbers 1, 2, 3, etc. Each router is assigned one mini-slot, but one mini-slot may be assigned to different routers. The mini-slot assignment algorithm is presented in Sect. 6.2.2.

When a router (say router A with mini-slot k) has packet(s) to transmit, it first monitors the mini-slots from 1 to $k-1$. If a jamming signal is detected at any of the mini-slots, it gives up the transmission at the current slot. Otherwise (i.e., the channel remains idle, which means that all the routers within two hops from router A and associated with mini-slot 1 to $k-1$ have no packet to transmit), router A sends a jamming signal at mini-slot k. By adjusting the transmission power of the jamming signals and the receivers' sensitivity, we can ensure that all the routers within two hops from router A hear the jamming signal.[2] Consequently, all of the one-hop and two-hop neighbors of router A will not transmit at the current slot to avoid corrupting router A's transmission. The router which sends a jamming signal at the control part will transmit its packets at the transmission part of the same slot.

6.2.2 Mini-slot Assignment

The mini-slot assignment has the following requirements: (1) Any two routers which are within the two-hop neighborhood of each other should not be assigned the same mini-slot; (2) A minimum number of mini-slots should be assigned. In other words, the number of mini-slots cannot be reduced without violating requirement (1). The first requirement is to ensure that the routers which send jamming signals at the same mini-slot can transmit simultaneously without interfering with each other. The second requirement is to reduce the control overhead as much as possible. A mini-slot assignment algorithm which satisfies these two requirements is proposed in the following. Since the routers are located at fixed sites, the mini-slot assignment can be determined based on the whole network topology at the initialization of the network.

The overhead of the proposed scheme is dependent on the maximal number of routers in a two-hop neighborhood but not the total number of routers in the network, making the proposed scheme scalable for large networks. Since the overhead caused by mini-slots in our scheme is much smaller than that caused by the backoff and RTS/CTS control message exchanging in contention-based schemes, the control overhead in the proposed scheme is expected to be greatly reduced.

6.2.3 Maximal Spatial Frequency Reuse

The proposed scheme can achieve maximal spatial frequency reuse. By maximal spatial frequency reuse we mean that the set of routers which transmit simultaneously (without interfering with each other) in each slot is a maximal set. That is, there does not exist any router which does not belong to this set but can transmit simultaneously (without interfering with each other) with all the routers in the set.

---

[2]Here we consider a good propagation environment. When router A sends a jamming signal, it is possible that some of its two-hop neighbors may not hear the jamming signal if there are obstacles in between. In this case, we let each router send jamming signals to its one-hop neighbors (with lower power), and split one mini-slot into two parts. In the first part, router A sends a jamming signal to its one-hop neighbors. Upon hearing the jamming signal, all its one-hop neighbors relay the jamming signal in the second part. Therefore, all the two-hop neighbors of router A can hear the jamming signal.

1: $N_m = 1$;   //N_m denotes the number of mini-slots. At the beginning of the algorithm, it
   is set to 1.
2: $S = \{$all the routers in the networks$\}$, $S_1 = NULL$;   //S_i denotes the set of routers which
   are assigned mini-slot i.
3: **while** $S \neq NULL$ **do**
4:       Randomly choose a router (denoted by A) from S
5:       assign_flag = FALSE
6:       **for** $i = 1,..,N_m$ **do**
7:            **if** none of one-hop and two-hop neighbors of router A belongs to S_i **then**
8:                 Assign mini-slot i to router A, and add router A into S_i;
9:                 Delete router A from S;
10:                assign_flag = TRUE;
11:                break;
12:           **end if**
13:       **end for**
14:       **if** assign_flag = FALSE **then**
15:            $N_m = N_m + 1$;
16:            Assign mini-slot N_m to router A, $S_{N_m} = \{A\}$;
17:            Delete router A from S;
18:       **end if**
19: **end while**

Algorithm 1: Mini-slot assignment

Proof. Consider a slot T. Let S_T denote the set of routers which transmit at slot
T. Suppose there exists one router A which does not belong to S_T (i.e., does not
transmit at slot T), and whose potential transmission at slot T does not interfere
with the transmissions of all the routers in S_T. Router A does not transmit at slot
T means that router A hears the jamming signal from one router (say B) within its
two-hop neighborhood. Thus, router B must be in S_T. Since router B is within the
two-hop neighborhood of router A, a collision can happen if both A and B transmit
to a same neighbor. This conflicts with the supposition.

6.2.4 Per-Router Fairness and Per-Flow Fairness

Two fairness models are considered: per-router fairness and per-flow fairness. In
the per-router fairness model, all the routers have fair channel access opportunities
independent of the number of micro-flows delivered by the routers. Thus, the flows
may have different throughput, depending on the traffic load of the associated
routers. In the per-flow fairness model, when any two routers (which may relay
different numbers of flows) contending with each other, all the flows[3] relayed by
the two routers have fair channel access opportunities. Thus, a heavy-load router
should have more chances to access the channel than a router with light load.

---

[3]Note that the flow here is not referred to as the end-to-end multi-hop flow, but the one-hop sub-
flow from the relay router to the next hop.

First, we consider how to achieve per-router fairness. From Sect. 6.2.1, it is obvious that the opportunity that one router may transmit in a slot largely depends on its mini-slot index in that slot. The smaller the index, the larger the opportunity. In order to fairly allocate the slots to each router, we have an initial mini-slot assignment (pre-determined at the initialization of the network), and rotate the index of the mini-slots slot by slot (i.e., the first mini-slot in the current slot becomes the last one in the next slot, the second mini-slot in the current slot becomes the first one in the next slot, and so on). It is possible that some routers may have less neighbors than others, i.e., the number of neighbors (within two-hop vicinity) of a router may be less than the number of mini-slots. In this case, just rotating the mini-slots may not ensure fair channel access for each router. Consider an example that a router (denoted by A) has 3 one-hop and two-hop neighbors B, C, and D, while the number of mini-slots is 6. A possible mini-slot assignment is shown in Fig. 6.2b. Accordingly, when we rotate the mini-slots, router A gets more chances to access the channel, benefiting from the two idle mini-slots. To solve this problem, we do not use a fixed mini-slot assignment. After a certain period, the order of the mini-slots is re-arranged (e.g., router D is assigned the first mini-slot and router A is assigned the 4th mini-slot), and each router rotates the mini-slots based on the new mini-slot assignment. All the mini-slot assignments are pre-determined and known by all the routers.

Per-flow fairness is achieved based on per-router fairness. Each router needs to exchange the information (i.e., the number of flows relayed by each router) with its one-hop and two-hop neighbors. According to the information, each router determines the fraction of time that it accesses the channel. Then each router adjusts its channel contending behavior accordingly. Consider an example that three routers (A, B, and C) contend with each other, while router A has 1 flow, router B has 2 flows, and router C has 3 flows. According to per-flow fairness, the fractions of channel time allocated to routers A, B, and C are $1/6$, $2/6$, and $3/6$, respectively. With per-router fairness, all the fractions of channel access time of the three routers are $1/3$. For router A, to reduce its time fraction from $1/3$ to $1/6$, it gives up half of its transmission opportunities. Thus, every two times when router A gets a turn to send a jamming signal at mini-slot 1, it gives up sending the jamming signal one time. On the contrary, to increase the time fraction of router C from $1/3$ to $3/6$, router C takes advantage of the transmission chances given up by router A. Router C can send its jamming signal at its own mini-slot upon hearing an idle channel during all the prior mini-slots. For router B, it neither gives up its own transmission opportunities nor takes the chances from others, thus maintains the same time fraction as that in per-router fairness. When the mini-slot of router B is not the last one, after hearing an idle channel during all the prior mini-slots, it does not send its jamming signal and leave the chance to router C. However, if the mini-slot of router B is the last one, it will transmit at the current slot to achieve spatial frequency reuse.

6.2.5   Guaranteed Priority Access for Real-Time Traffic

Since real-time traffic usually has a strict delay requirement, guaranteed priority access for real-time traffic is necessary in order to provide QoS satisfaction for real-time traffic. Hence, we add a mini-slot prior to all the other mini-slots. This extra mini-slot (referred to as real-time mini-slot) is dedicated to real-time traffic and its index is not rotated. For a router with real-time packet(s) to transmit, in addition to sending a jamming signal in its own mini-slot, it first sends a jamming signal in this real-time mini-slot. Upon hearing the jamming signal in this mini-slot, the routers which have only data packets will not send their own jamming signals, letting routers with real-time traffic send a jamming signal in their corresponding mini-slots. When two or more real-time routers contend for the same slot, the one with the smallest mini-slot index will first send the jamming signal and get the slot.

In order to provide further priority differentiation to real-time packets with different delay requirements, we can have an additional number of real-time mini-slots. For real-time mini-slot i ($i = 1, 2, \ldots$), a corresponding urgency level U_i ($U_1 < U_2 < U_3 \ldots$) is pre-defined. The smaller the U_i, the more urgent the level is. The urgency of a real-time packet is measured by the packet due time and the remaining hops to the destination. The due time of a real-time packet is the packet generation time plus the packet delay bound. We assume that this information is included in the packet header and known by the traversed routers. The packet is more urgent if the due time is smaller and the number of the remaining hops is larger. Considering a router with a real-time packet j having the remaining time t_j to the due time and the remaining hops n_j to the destination, if $U_{i-1} < \frac{t_j}{n_j} \leq U_i$ (where $U_0 = 0$), then the router sends a jamming signal at real-time mini-slot i if all the prior real-time mini-slots are idle. Once a router hears the jamming signal (which means that another one-hop or two-hop-away router has a more urgent real-time packet), it will quit the contention for the current slot.

6.2.6   Congestion Avoidance

In the wireless backbone, it is very likely that some routers (referred to as bottleneck routers) located at the center of the network or near the gateway need to relay more traffic than other routers. In the case of per-router fairness, with an absolute fair channel access for each router, the traffic arrival rate will be higher than the traffic departure rate at the bottleneck routers. As a result, the packets will be accumulated, eventually causing buffer overflow at the bottleneck routers. It is possible that multi-hop data flows pass through bottleneck routers. Buffer overflow at the routers results in resource waste and low end-to-end throughput. TCP is the most popular protocol to deal with network congestion at the transport layer. However, TCP suffers from severe performance degradation in wireless networks, due to the fact that it is difficult for the source nodes at the transport layer to know explicitly whether a packet loss is due to buffer overflow or temporary link failure [35]. In order to avoid congestion effectively in the mesh backbone, we propose a straightforward mechanism at the MAC layer. Each router keeps track of its packet arrivals and departures. For each one-hop neighbor, the router records the number of arrived packets (denoted by C_a) and departed packets (denoted by C_d) which are from the neighbor. If the difference between C_a and C_d is larger than a pre-defined threshold, the router sends a message to the neighbor to suspend its transmissions to this router. When the difference between C_a and C_d decreases to a certain value, the router sends a message to resume the transmissions. This approach avoids buffer overflow at intermediate hops of multi-hop flows, in order to more efficiently utilize network resources for a higher end-to-end throughput. The control propagates hop by hop to the source node and regulates the source rate depending on the network congestion status.

6.3 Performance Analysis

To make the analysis tractable, we consider a simplified case that (1) there is one real-time mini-slot, and all the real-time packets are treated equally; (2) per-router fairness is considered. We assume that the voice and video call arrivals at each source node are independent and follow a Poisson process, and the call duration has an exponential distribution.

6.3.1 Real-Time Traffic Access Delay Bound

The access delay is defined as the time period from the instant that a packet becomes the head in the buffer to the instant that the packet departs from the router. Let T_s denote the time duration of one slot, and N^m the number of mini-slots, including the real-time mini-slot. Consider the worst case that the target router has $N^m - 2$ one-hop and two-hop neighbors, and all of them have real-time packets to transmit. After the target router transmits one packet, it takes $T_s \cdot (N^m - 1)$ for the target router to transmit the next one. Thus, the access delay bound of real-time traffic at each hop is $T_s \cdot (N^m - 1)$, which is independent to the traffic load of the networks.

6.3.2 Data Traffic Access Delay

Since guaranteed priority access is provided to real-time traffic, the real-time traffic load will impact the data traffic access delay. The voice call is represented by an on/off model with parameters α and β. At an on state, voice packets are generated periodically with an inter-arrival time I_o, while no voice packet

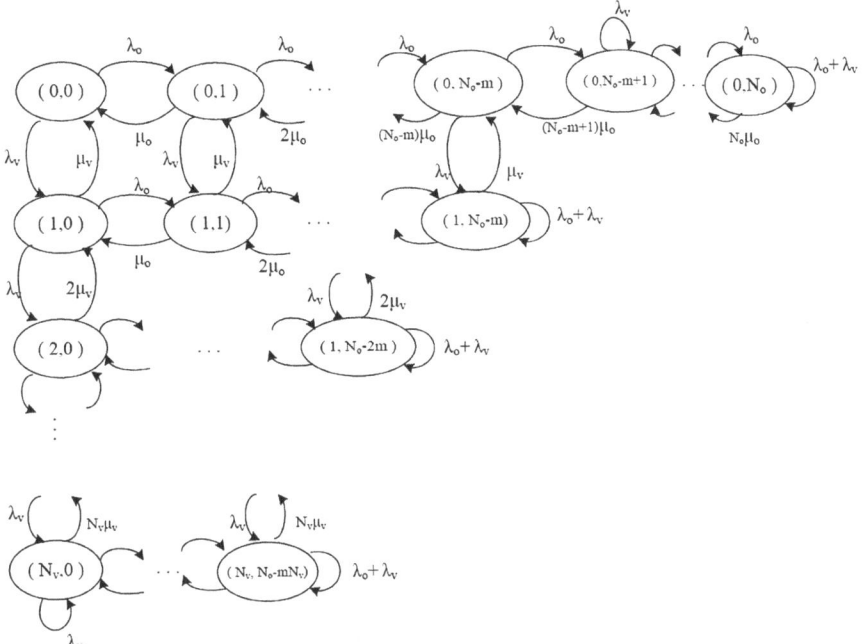

Fig. 6.3 The state transition diagram of (n_v, n_o)

is generated at an `off` state. For a video call, the video frames are generated periodically with an inter-arrival time I_v. The video frame usually has a large and variable size [93]. Suppose that it takes one slot to transmit one voice packet, and M_v slots (on average) to transmit one video frame. Considering a target router, we refer to its two-hop vicinity as the target area.

To obtain the data traffic access delay, we first need to derive the fraction of channel time occupied by voice and video traffic. We define a two-dimensional state (n_v, n_o), where n_v and n_o are the numbers of video calls and voice calls, respectively, being served by the routers within the target area. Denote the average arrival rates of voice and video calls that traverse the target area as λ_o and λ_v, respectively, and the average call duration as μ_o^{-1} and μ_v^{-1}, respectively. We assume that call admission control is in place to guarantee the QoS of voice and video calls, and the maximal number of acceptable voice and video calls within the target area are denoted by N_o and N_v, respectively. The state transition diagram is shown in Fig. 6.3. Since a video call requires more resources than a voice call, when there is 1 video call being served, the maximal number of supported voice calls is $N_o - M_v \cdot \frac{I_o}{I_v}$, denoted by $N_o - m$. Define p_{ij} as the joint probability that i video calls and j voice calls being served. The balance equations for the two-dimensional state space of Fig. 6.3 are

$i = 0, j = 0:$ $(\lambda_o + \lambda_v)p_{00} = \mu_o p_{01} + \mu_v p_{10};$

$i = 0, 1 \leq j \leq N_o - m:$ $(\lambda_o + \lambda_v + j\mu_o)p_{0j} = \lambda_o p_{0j-1}$

$$+ (j+1)\mu_o p_{0j+1} + \mu_v p_{1j};$$

$i = 0, N_o - m + 1 \leq j \leq N_o - 1:$ $(\lambda_o + j\mu_o)p_{0j} = \lambda_o p_{0j-1} + (j+1)\mu_o p_{0j+1};$

$i = 0, j = N_o:$ $N_o \mu_o p_{0N_o} = \lambda_o p_{0N_o-1};$

$1 \leq i \leq N_v - 1, j = 0:$ $(\lambda_o + \lambda_v + i\mu_v)p_{i0} = \lambda_v p_{i-10}$

$$+ (i+1)\mu_v p_{i+10} + \mu_o p_{i1};$$

$1 \leq i \leq N_v - 1,$

$1 \leq j \leq N_o - m(i+1):$ $(\lambda_o + \lambda_v + i\mu_v + j\mu_o)p_{ij} = \lambda_v p_{i-1j} + \lambda_o p_{ij-1}$

$$+ (j+1)\mu_o p_{ij+1} + (i+1)\mu_v p_{i+1j};$$

$1 \leq i \leq N_v - 1,$

$N_o - m(i+1) + 1 \leq j \leq N_o - im - 1:$ $(\lambda_o + i\mu_v + j\mu_o)p_{ij} = \lambda_v p_{i-1j} + \lambda_o p_{ij-1}$

$$+ (j+1)\mu_o p_{ij+1};$$

$1 \leq i \leq N_v - 1, j = N_o - im:$ $(j\mu_o + i\mu_v)p_{ij} = \lambda_v p_{i-1j} + \lambda_o p_{ij-1}.$

For the case of $i = N_v$, we need to consider three possibilities: $N_o - mN_v > 1$, $N_o - mN_v = 1$, and $N_o - mN_v = 0$. When $N_o - mN_v > 1$, the corresponding balance equations are

$i = N_v, j = 0:$ $(\lambda_o + i\mu_v)p_{ij} = \lambda_v p_{i-1j} + \mu_o p_{ij+1};$

$i = N_v, 1 \leq j \leq N_o - mN_v - 1:$ $(\lambda_o + i\mu_v + j\mu_o)p_{ij} = \lambda_v p_{i-1j} + \lambda_o p_{ij-1}$

$$+ (j+1)\mu_o p_{ij+1};$$

$i = N_v, j = N_o - mN_v:$ $(i\mu_v + j\mu_o)p_{ij} = \lambda_v p_{i-1j} + \lambda_o p_{ij-1}.$

When $N_o - mN_v = 1$, the corresponding balance equations are

$i = N_v, j = 0: (\lambda_o + i\mu_v)p_{ij} = \lambda_v p_{i-1j} + \mu_o p_{ij+1};$

$i = N_v, j = 1: (i\mu_v + \mu_o)p_{ij} = \lambda_v p_{i-1j} + \lambda_o p_{ij-1}.$

When $N_o - mN_v = 0$, the corresponding balance equations is

$$i = N_v, j = 0: \quad i\mu_v p_{i0} = \lambda_v p_{i-10}.$$

Based on the above balance equations, the probability distribution of state (n_v, n_o) can be derived. A voice/video call may traverse several hops within the target area. Let h_n^o and h_n^v denote the average number of hops that voice and video calls traverse the target area, respectively. As voice traffic only generates packets during an on period, at any time instant, each voice call is at the on state with probability $\beta/(\alpha + \beta)$. During I_o (i.e., the voice packet inter-arrival duration), each voice call which is at the on state generates one voice packet. Thus, given n_o voice calls being served in the target area, the average channel time occupied by these n_o voice calls during I_o is given by

$$\overline{T_o}(n_o) = \sum_{i=1}^{n_o} \binom{n_o}{i} \left(\frac{\beta}{\alpha + \beta}\right)^i \left(\frac{\alpha}{\alpha + \beta}\right)^{n_o - i} \cdot i \cdot T_s \cdot h_n^o. \tag{6.1}$$

For a video call with the frame inter-arrival duration I_v, the average number of video frames that a video call generates during I_o is I_o/I_v. Thus, given n_v video calls, the average channel time occupied by these n_v video call during I_o is given by

$$\overline{T_v}(n_v) = \frac{I_o}{I_v} \cdot n_v \cdot M_v \cdot T_s \cdot h_n^v. \tag{6.2}$$

Thus, the fraction of channel time occupied by real-time traffic is given by

$$f = \sum_{\text{all state } (n_v, n_o)} \frac{(\overline{T_o}(n_o) + \overline{T_v}(n_v)) \cdot p_{n_v n_o}}{I_o}. \tag{6.3}$$

In our scheme, the residual channel time left by real-time traffic is fairly shared by all the routers with data traffic. For data traffic access delay, we consider two cases: saturated case and unsaturated case. In the saturated case, all the routers with data traffic always have data packets to transmit. In the unsaturated case, the average date packet arrival rate at each router is denoted as λ_d. First consider the saturated case. Given an arbitrary time slot, the probability that the target router can transmit its data packet in that slot is given by

$$p_t = (1 - f) \cdot \frac{1}{K} \tag{6.4}$$

where f is given by (6.3), and K is the number of data routers within the target area that fairly share the residual channel time left by the real-time traffic. Thus, the data traffic access delay of the target router is $\frac{T_s}{p}$. For the unsaturated case, denote the data traffic access delay as d_a. The data packet arrivals and departures at each router can be considered as a queue, and the queue utilization is $\rho = \lambda_d d_a$ ($\rho \le 1$). For each router with data traffic, at an arbitrary time, the router has data packet(s) to transmit with probability ρ and has no data packet to transmit with probability $1 - \rho$. Thus, (6.4) can be re-written as

$$p_t = (1 - f) \cdot \sum_{i=0}^{K-1} \binom{K-1}{i} \cdot \rho^{K-1-i} \cdot (1 - \rho)^i \cdot \frac{1}{K - i}, \quad \rho \le 1. \tag{6.5}$$

Substituting $p_t = \frac{T_s}{d_a}$ and $\rho = \lambda_d d_a$ in (6.5), we can obtain d_a. Note that when $\rho = 1$ (i.e., the saturated case), (6.5) is equivalent to (6.4).

6.3.3   Numerical Results

Simulations are carried out in order to verify the accuracy of the analysis. Since the analysis of real-time access delay bound is straightforward, here we validate

Table 6.1 The average data traffic access delay (ms) with different λ_v (call/s) while $\lambda_o = 0.1$ call/s

Video call arrival rate λ_v		0.01	0.025	0.05	0.075	0.1
Data access delay	Simulation	7.99	12.47	16.76	20.56	26.55
	Analysis	7.09	11.42	16.67	21.62	26.91

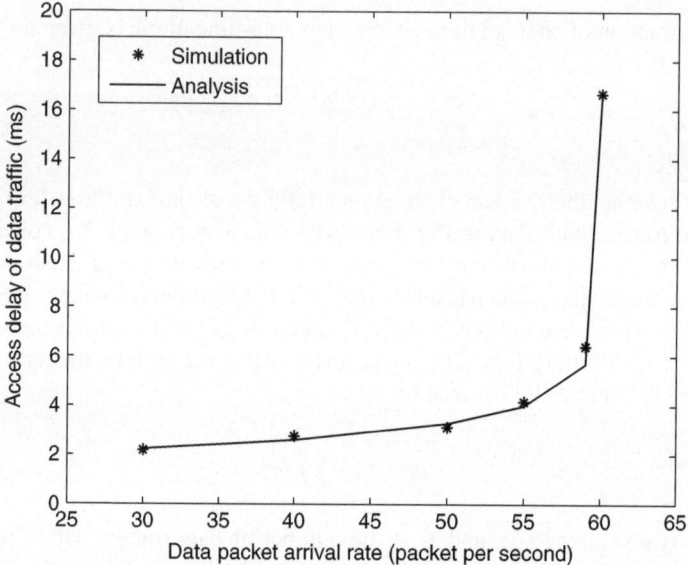

Fig. 6.4 The data traffic access delay with different data packet arrival rate

the analysis of data traffic access delay. Without loss of generality, we choose the parameters $T_s = 0.2$ ms,[4] $K = 10$, $N_o = 40$, $N_v = 5$, $h_n^o = 3$, and $h_n^v = 3$. The voice packet and video frame inter-arrival durations are 20 and 100 ms, respectively. One video frame takes 40 slots (on average) to transmit. The average voice and video call durations are 150 and 600 s, respectively. For a voice call, the average on and off durations are 352 and 650 ms, respectively.

First, consider the saturated case. We fix the voice call arrival rate λ_o as 0.1 call/s, and vary the video call arrival rate λ_v from 0.01 to 0.1 call/s. Table 6.1 compares the simulation and analytical results of the data traffic access delay. They agree with each other well.

Second, consider the unsaturated case. We fix λ_o and λ_v as 0.1 and 0.05 call/s, respectively, and vary the average date packet arrival rate λ_d from 30 to 60 packet/s. Figure 6.4 shows the data traffic access delay. Note that when $\lambda_d = 60$ packet/s, ρ equals to 1 (i.e., $\lambda_d d_a = 1$), the data routers become saturated. The data access delay

---

[4]According to [24, 29, 77], the current synchronization technology can achieve synchronization accuracy of less than 20 μs.

increases sharply when the system approaches the saturated case. It is clear that the simulation results match well with the analytical results.

6.4   Performance Evaluation

We evaluate the performance of the proposed scheme by extensive simulations. For comparison, we demonstrate the performance of IEEE 802.11 as well. Although IEEE 802.11 MAC is not designed for wireless mesh networks, here we compare our scheme with IEEE 802.11 because it is the most popular distributed MAC scheme and there is no representative distributed MAC scheme for wireless mesh networks so far. For voice traffic, we choose the GSM 6.10 codec as an example. For video traffic, we choose the H.264 codec, which is the most efficient video compression technology and is widely implemented. The H.264 defines a set of profiles with different video bit rates for various classes of applications. Here, we use H.264 with video bit rate of 384 kbps. The frame rate is 30 frame/s. For data traffic, the data packet arrivals follow a Poisson process with various arrival rates. Other simulation parameters follow the IEEE 802.11g/e standards [2, 3], where the channel rate (i.e., 54 Mbps) is to transmit voice/video/data packets, and the basic rate (i.e., 24 Mbps) is to transmit RTS and CTS (in IEEE 802.11). The threshold for congestion control (i.e., $C_a - C_d$) is chosen to be 5 packets. A router will suspend a neighbor's transmission if the value of $C_a - C_d$ of that neighbor reaches 5. When the value decreases to zero, the router will resume the suspended transmission of that neighbor. From the simulations, we observe that our scheme is not sensitive to the value of the threshold as long as it is not very large. Varying the threshold from 5 packets to 50 packets does not result in much differences in the following simulation results.

6.4.1   The Delay Performance for Real-Time Traffic

We consider the case that there is one real-time mini-slot in our scheme, and voice traffic and video traffic are treated equally. We use video traffic as an example to demonstrate the performance of guaranteed priority access. A chain topology as shown in Fig. 6.5a is considered. We consider two video flows, flow 1 having 4 hops from router 1 to the gateway, and flow 2 having one hop from router 4 to the gateway. Each flow is an aggregated video flow, including 10 video calls. To demonstrate the performance of priority access for real-time traffic, we let the two video flow experience various contention degrees with data traffic. First, consider the case that there is no other data flow contending with these two video flows. Then we increase the number of data contenders N_{dc} near each router gradually from 1 to 5, each contender having a data flow with source rate 3 Mbps to the gateway.

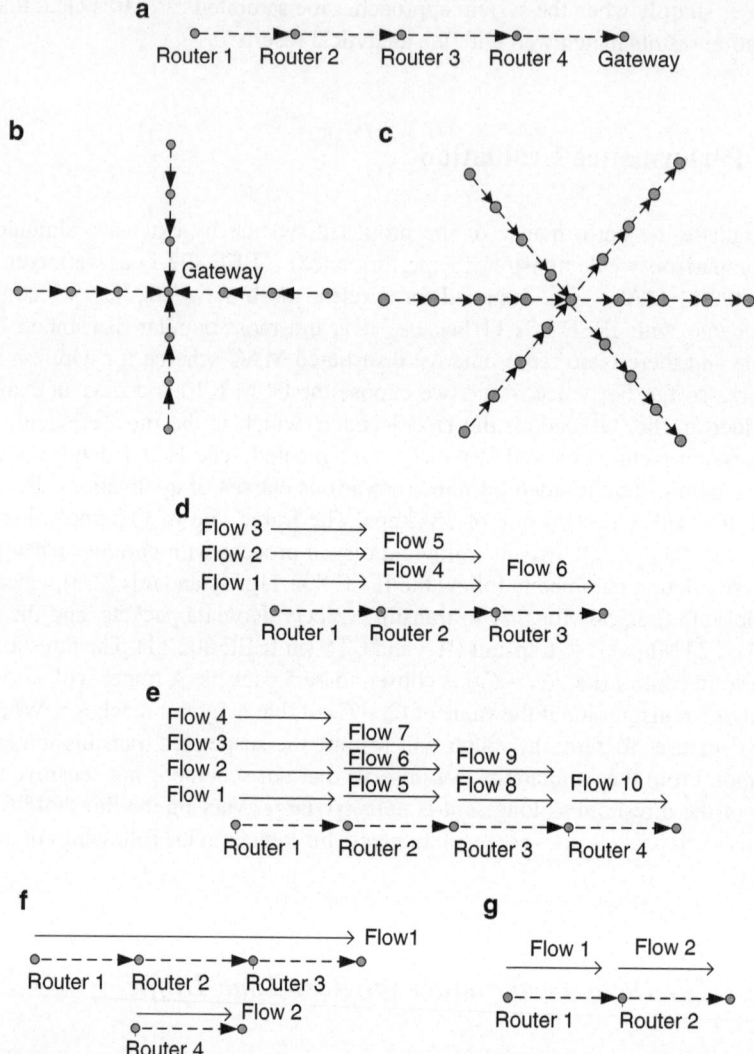

Fig. 6.5 The simulation topologies used in performance evaluation of the proposed collision-free MAC scheme

Table 6.2 compares the packet delay of the two video flows. It can be seen that, as the number of data contenders increases, the video packet delays increase in IEEE 802.11. Especially for flow 1 (with a relatively long path), the video packet delay increases significantly from 2.12 to 128.24 ms when the number of data contenders varies from 0 to 5. On the contrary, in our scheme, the video packet delays remain stable for all the numbers of data contenders. These results demonstrate that in IEEE 802.11, the delay performance of real-time traffic is degraded when the

Table 6.2 The average video packet delay (ms) with different number of data contenders near each router

N_{dc}		0	1	2	3	4	5
802.11	Flow 1	2.12	4.01	26.02	63.96	93.55	128.24
	Flow 2	0.44	0.75	1.88	1.96	2.73	4.40
Proposed	Flow 1	0.67	0.67	0.67	0.68	0.67	0.68
	Flow 2	0.21	0.21	0.21	0.21	0.21	0.21

data traffic load increases. Especially for real-time flows with a long path, such performance degradation is significant. The reason is that IEEE 802.11 provides statistical priority access, which is difficult to satisfy the delay requirement of real-time traffic since the real-time traffic may suffer from performance degradation due to a high data traffic load [72]. Video flows with a long path also suffer from the priority reversal problem, resulting in even worse delay performance. Contrary to IEEE 802.11, our scheme can achieve guaranteed priority access for real-time traffic. In addition, as the routers within the two-hop neighborhood are not allowed to transmit simultaneously, hidden terminals do not exist, neither does the priority reversal problem. As a result, our scheme achieves a small delay for both long-path and short-path real-time flows regardless of the data traffic load.

6.4.2 Fairness and End-to-End Throughput of Data Flows

For data flows, the QoS metrics of interest are fairness and end-to-end throughput. Here, we consider two scenarios: the chain topology in Fig. 6.5a with 4 data flows, where flow i ($i = 1, \ldots, 4$) is from router i ($i = 1, \ldots, 4$) to the gateway; and the cross topology with 12 data flows in Fig. 6.5b. In the cross topology, the center node is the gateway, and each router has a data flow to the gateway.

For the chain topology, we vary the data packet arrival rate of each flow, and obtain the end-to-end throughput of the 4 flows, given in Table 6.3. The end-to-end throughput is measured by the number of data packets received at the gateway. For IEEE 802.11, all the flows have the same throughput when the traffic load is low. However, when the traffic load becomes high, the resources are not fairly allocated to each flow. The throughput of the flow with the shortest path (i.e., flow 4) is much larger than that of the flow with the longest path (i.e., flow 1). The unfairness is due to the hidden terminal problem, as discussed in Chap. 2. However, the throughputs of the 4 flows in our scheme are always the same under the varying traffic loads, indicating that our scheme has improved fairness performance over IEEE 802.11. In addition, the aggregated throughput in our scheme is always larger than that in IEEE 802.11. By avoiding collisions and reducing the control overhead, our scheme achieves higher resource utilization than IEEE 802.11. Similar results are observed in the cross topology, as shown in Table 6.3. In contrast to IEEE 802.11 that the

Table 6.3 The end-to-end throughputs (Mbps) of data flows in the chain topology and cross topology

Source rate of each flow (Mbps)			1	3	5	7	9
		Flow 1	0.97	0.82	0.53	0.23	0.19
		Flow 2	0.97	0.98	0.59	0.54	0.26
	802.11	Flow 3	0.97	2.83	2.59	2.48	2.14
		Flow 4	0.97	2.86	4.42	5.91	6.78
		Aggregate	3.88	7.49	7.83	9.16	9.37
The chain topology		Flow 1	0.98	2.98	4.40	4.42	4.42
		Flow 2	0.99	2.99	4.41	4.42	4.43
	Proposed	Flow 3	0.99	2.99	4.42	4.43	4.42
		Flow 4	0.98	2.98	4.44	4.44	4.44
		Aggregate	3.94	11.94	17.67	17.71	17.71
Source rate of each flow (Mbps)			0.5	1	2	3	4
		One-hop flow	0.49	0.95	1.63	1.92	1.49
		Two-hop flow	0.49	0.80	0.54	0.39	0.30
	802.11	Three-hop flow	0.48	0.79	0.53	0.38	0.29
		Aggregate	5.84	10.16	10.80	10.76	8.32
The cross topology		One-hop flow	0.49	0.99	1.99	2.98	3.03
		Two-hop flow	0.49	0.99	1.99	2.98	3.02
	Proposed	Three-hop flow	0.49	0.99	1.99	2.98	3.06
		Aggregate	5.88	11.88	23.88	35.76	36.44

one-hop flows achieve a much higher throughput than the two-hop and three-hop flows when the traffic load increases, in our scheme, all the flows have almost the same throughput in all the cases.

6.4.3   Relay Efficiency

As mentioned earlier, if a MAC scheme is designed without considering congestion avoidance, it is very likely that the source nodes may inject more packets than what the bottleneck routers can forward. As a result, some packets sent by the source nodes are dropped by the bottleneck routers due to buffer overflow, leading to a waste of wireless channel and power resources. Relay efficiency is defined as the ratio of the sum of the packets received at all the destinations to the sum of the packets sent by all the sources. This metric reflects how much the resources are wasted in relaying. The smaller the relay efficiency, the more the resources are wasted at the bottleneck routers. The scenario of three data flows as shown in Fig. 6.5c is considered, where the router at the center is a bottleneck router, relaying all the data flows.

In Fig. 6.6, we show the relay efficiency versus the offered traffic load. It is clear that our scheme achieves close to 100 % relay efficiency in all the cases, while IEEE

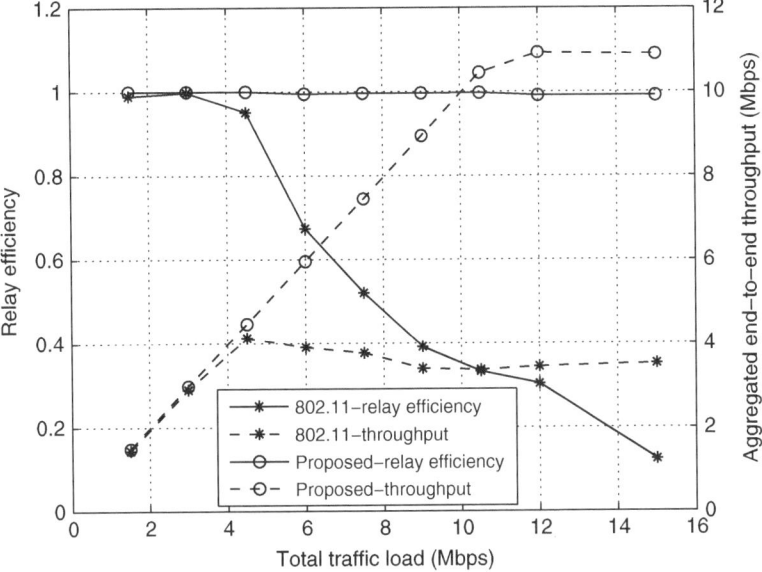

Fig. 6.6 The relay efficiency and aggregated end-to-end throughput in the cross topology

802.11 has a decreased efficiency when the traffic load increases. The result implies that, without congestion control, IEEE 802.11 drops more and more packets at the bottleneck router, as the traffic load increases. For comparison, the aggregated end-to-end throughputs of the two schemes are also shown in Fig. 6.6. By avoding packet dropping at the bottleneck router, our scheme utilizes the resources more efficiently and achieves a higher end-to-end throughput. Note that the relay efficiency is close to 100 % in our scheme, so the packet sending rate at the source nodes almost equals to the packet receiving rate at the destinations (i.e., the end-to-end throughput). We notice that, in our scheme, the aggregated packet sending rate at all the source nodes is bounded at about 11 Mbps although the total traffic load (i.e., the aggregated packet arrival rate at all the source nodes) may be higher than 11 Mbps. This result indicates that our scheme effectively controls the source sending rate, so that congestion can be avoided.

6.4.4 Performance in Random Topology

To evaluate the performance of the propose scheme in a more general case, we consider a random topology. In the simulations, 100 routers are uniformly placed deterministically in a $1,000 \times 1,000 \, m^2$ area. Two routers are one-hop neighbors if the distance between them is less than or equal to $100 \, m$. Fifty flows are considered, with 5 voice flows, 5 video flows, and 40 data flows. The source and destination

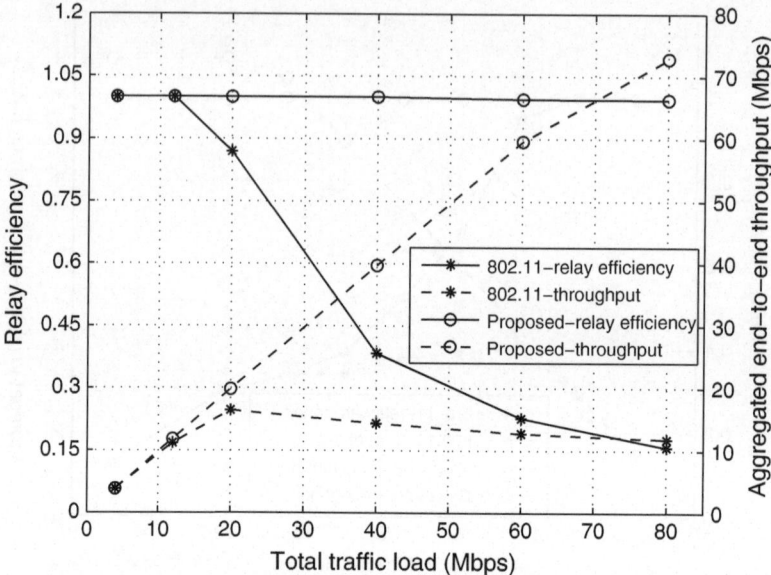

Fig. 6.7 The relay efficiency and aggregated end-to-end throughput in the random topology

of each flow are randomly selected, and a shortest path from the source to the destination is pre-determined, so that each intermediate router knows its up-stream and down-stream routers. We vary the traffic load of each data flow from 0.1 to 2 Mbps and observe that the end-to-end delays of one randomly picked voice and video flow remain unchanged at 0.63 and 1.46 ms, respectively, over the data traffic load range. This result confirms again that our scheme provides guaranteed priority access to real-time traffic regardless of data traffic load.

From Fig. 6.7, we can see that the relay efficiency of our scheme is almost 100 % under all the cases of data traffic load. However, in IEEE 802.11, the relay efficiency drops rapidly when the data traffic load increases. Figure 6.7 also compares the aggregated end-to-end throughput of the 40 data flows in IEEE 802.11 and in our scheme. When the traffic load increases, our scheme achieves a much higher end-to-end throughput than IEEE 802.11.

We use Jain's Fairness Index [40] to investigate the fairness performance. Figure 6.8 compares the Fairness Index values of our scheme and IEEE 802.11. When the total traffic load is low (i.e., less than 12 Mbps), the Fairness Index values of both schemes are 1, while with the increase of traffic load, our scheme achieves much better fairness performance than IEEE 802.11. Note that when the traffic load is at 80 Mbps, the Fairnesses Index value in our scheme is less than 1. Since flows are randomly chosen, some flows may experience more contentions than others. Due to the capacity limit, the flows with more contentions cannot further increase their throughput, while other flows with less contentions may still increase their throughput.

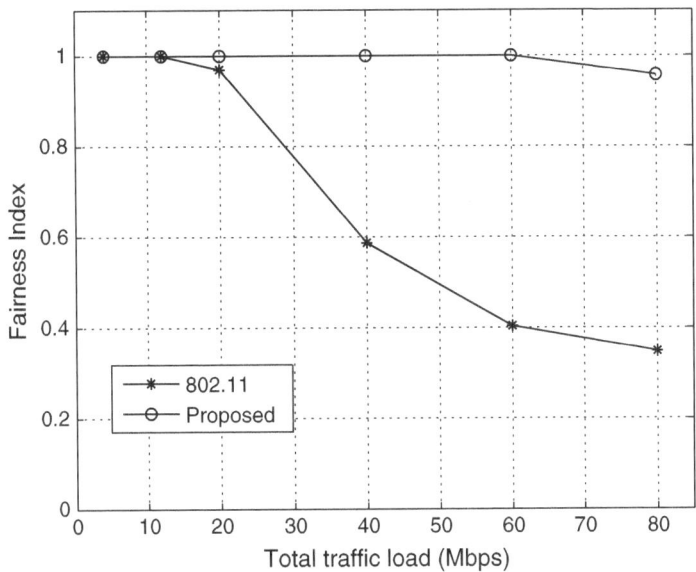

Fig. 6.8 The Fairness Index in the random topology

6.4.5 The Comparison of Per-Flow Fairness and Per-Router Fairness

In the preceding simulations, one router generates one data flow and per-router fairness is considered. In the following, we consider the cases that different routers generate different numbers of data flows and compare the performance of per-router fairness and per-flow fairness. First, consider the scenario shown in Fig. 6.5d, where there are 6 data flows contending with each other. We vary the data packet arrival rate of each flow, and obtain the throughput of each flow under per-flow fairness and per-router fairness, respectively, given in Table 6.4. It can be seen that, in the case of per-router fairness, all the flows have the same throughput when the traffic load is low. When the traffic load becomes high, flow 6 has the highest throughput, and flows 1–3 have the lowest throughput. In the case of per-flow fairness, all the flows have the same throughput in all the cases.

Second, consider the scenario shown in Fig. 6.5e, which has 10 data flows. Similar observation can be found in Table 6.4. Note that in both per-flow and per-router fairness cases, flow 10 has a much higher throughput than other flows. It is because router 4 can transmit simultaneously with router 1 for spatial frequency reuse. Also note that the aggregate throughput of all the flows with per-flow fairness is lower than that with per-router fairness. In the case of per-flow fairness, each router exchanges the flow information only with its one-hop and two-hop neighbors, and adjusts its channel access time accordingly. Due to the lack of the flow information of the whole network, the resources may not be fully utilized.

Table 6.4 The throughputs (Mbps) of the data flows in scenarios shown in Fig. 6.5d, e

Source rate of each flow (Mbps)			4	6	8	10	15
		Flow 1–3	3.99	5.98	5.29	5.00	4.43
	Per-router fairness	Flow 4–5	3.99	5.98	7.98	7.44	6.65
		Flow 6	3.99	5.98	7.99	9.99	13.29
Scenario (d)		Flow 1–3	3.99	5.98	6.64	6.64	6.64
	Per-flow fairness	Flow 4–5	3.99	5.98	6.65	6.65	6.65
		Flow 6	3.99	5.98	6.65	6.65	6.65
Source rate of each flow (Mbps)			2	4	6	8	10
		Flow 1–4	1.98	3.98	3.34	3.32	3.32
		Flow 5–7	1.98	3.98	4.82	4.42	4.42
	Per-router fairness	Flow 8–9	1.98	3.98	5.98	6.64	6.64
		Flow 10	1.99	3.99	5.98	7.99	9.99
		Aggregate	19.81	39.81	45.76	47.81	49.81
Scenario (e)		Flow 1–4	1.98	3.96	3.97	3.98	3.98
		Flow 5–7	1.98	3.96	3.98	3.98	3.99
	Per-flow fairness	Flow 8–9	1.98	3.97	3.99	3.99	3.99
		Flow 10	1.99	3.99	5.99	6.64	6.65
		Aggregate	19.81	35.69	41.79	42.48	42.52

For example, after exchanging the flow information with routers 2 and 3, router 4 considers the fractions of channel access time of routers 2, 3, and itself are $1/2$, $1/3$, and $1/6$, respectively. However, router 2 can only get $1/3$ of channel time because of the contending flows from router 1. Without knowing this information, router 4 cannot fully utilize the channel time which is not used by router 2.

6.4.6 Priority Differentiation of Real-Time Packets

In the preceding simulations, the network has sufficient resources to transmit all the real-time traffic with no packet dropping, so all the real-time packets are treated equally. In the following, we consider two scenarios where the real-time traffic load exceeds the network capacity, leading to packet dropping. In these scenarios, further priority differentiation is needed. The first scenario is shown in Fig. 6.5f, where flows 1 and 2 both consist of 11 video calls, flow 1 having 3 hops and flow 2 having 1 hop. The number of real-time mini-slots n_r is chosen to be 10, and the video packet delay bound D_{max} is set as 100 ms. We consider two methods to differentiate the priorities. First, D_{max} is uniformly divided (we refer to this method as uniform priority differentiation), and the urgency level of real-time mini-slot i is given by $U_i = i \cdot \frac{D_{max}}{n_r}$, $(1 \le i \le n_r)$. Second, D_{max} is non-uniformly divided (we refer to it as non-uniform priority differentiation). For the packets which approach the due time, we differentiate them with a small scale. Those non-urgent packets are differentiated with a large scale. Specifically, the urgency level of real-time mini-slot i is given by

$$\begin{cases} U_i = \frac{1}{a} \cdot i \cdot \frac{D_{max}}{n_r}, & 1 \leq i \leq \lfloor \frac{n_r}{b} \rfloor \\ U_i = \frac{D_{max} - \frac{1}{a} \cdot \lfloor \frac{n_r}{b} \rfloor \cdot \frac{D_{max}}{n_r}}{n_r - \lfloor \frac{n_r}{b} \rfloor} \cdot (i - \lfloor \frac{n_r}{b} \rfloor) + \frac{1}{a} \cdot \lfloor \frac{n_r}{b} \rfloor \cdot \frac{D_{max}}{n_r}, & \lfloor \frac{n_r}{b} \rfloor + 1 \leq i \leq n_r, \end{cases}$$

where $\lfloor \cdot \rfloor$ is the floor function, a and b (both larger than 1) are the adjustable parameters. For the considered scenarios, when $a = 10, b = 2$, the desired priority differentiation performance is achieved. Without considering further priority differentiation, the packet dropping rates of flows 1 and 2 are 3.82 % and 0, respectively. With the uniform priority differentiation, they are 2.90 and 0.92 %, respectively. With the non-uniform method, they are 1.81 and 1.61 %, respectively. It is clear that without further priority differentiation, the packets are not fairly dropped, and all the packet dropping are from the flow with a relatively long path. The uniform priority differentiation is not effective to improve the fairness. With the non-uniform priority differentiation, the packets are dropped more or less fairly between the flows with different hops.

The second scenario is shown in Fig. 6.5g, where flows 1 and 2 consist of 40 and 10 video calls, respectively. Without considering further priority differentiation, the packet dropping rates of flows 1 and 2 are 3.05 % and 0, respectively. The packets from the heavy-load flows are more likely to be dropped. With non-uniform priority differentiation, the packet dropping rates of flows 1 and 2 are 2.13 and 2.10 %, respectively. The packets are dropped fairly between the flows regardless of the traffic load of each flow.

In addition to the methods discussed in this subsection, there are other different methods to provide priority differentiation of real-time packets, e.g., in [54, 55], where different weights are considered to balance the distance and the lifetime of packets when differentiating their priorities. Our scheme is not restricted to the preceding proposed methods. Other priority differentiation methods can also be adopted.

6.5 Summary

In this chapter, we propose a novel collision-free MAC scheme supporting multimedia traffic for the wireless mesh backbone. The proposed scheme is distributed, simple, and scalable. Taking the unique characteristics of the wireless mesh backbone into consideration, the proposed MAC greatly reduces the control overhead in comparison with conventional contention-based MAC schemes (e.g., IEEE 802.11). By eliminating collisions, reducing control overhead, and achieving maximal spatial frequency reuse, the proposed MAC achieves much higher resource utilization than contention-based MAC. In addition, the proposed scheme provides guaranteed priority access to real-time traffic and, at the same time, ensures fair channel access to data traffic. The simulation results demonstrate that it significantly improves the delay performance of real-time traffic, the fairness of data traffic, and the end-to-end data throughput, as compared with IEEE 802.11. The performance of the proposed scheme is analyzed and verified by computer simulations. This research should provide helpful insights to the development of future broadband wireless mesh networks.

Chapter 7
Conclusions

In this book, we have introduced several distributed MAC schemes to provide QoS support for different types of traffic in heterogeneous wireless networks including WLANs, wireless ad hoc networks, and wireless mesh networks. Specifically, these MAC schemes are summarized as follows:

- In Chap. 3, we propose mechanisms to enhance the voice QoS provisioning capability of infrastructure WLANs supporting hybrid voice/data traffic. Originally designed for high-rate data traffic, WLAN has limited capacity to support delay-sensitive voice traffic and may experience bandwidth inefficiency when supporting voice traffic. Previous work mostly focuses on contention-based medium access which cannot provide guaranteed QoS to voice traffic. Aiming at addressing these limitations, we combine the controlled access and the contention-based access to achieve voice traffic multiplexing, making use of the on/off characteristic of voice traffic. We also propose mechanisms to greatly reduce the overhead of voice traffic transmissions. By achieving voice traffic multiplexing and reducing overhead, the proposed scheme increases the voice capacity significantly, as compared with the current WLAN standard IEEE 802.11e. The voice capacity is also theoretically analyzed to facilitate call admission control. Our work provides helpful insights to the development and deployment of VoIP technologies over WLANs.
- In Chap. 4, we develop a novel token-based MAC scheme for ad hoc mode WLANs that supports both voice and data traffic. Most of the existing WLAN MAC schemes provide priority access by adjusting the contention window sizes and inter-frame spaces for different traffic classes. Although this method can provide a certain degree of service differentiation, it is difficult to quantify the degree of service differentiation, and even more difficult to adjust the degree flexibly among different classes based on some specific requirements of customers or network service providers. To address this limitation, a token-based MAC scheme is proposed which can provide more precise and quantitative service differentiation for data traffic and, at the same time, provide guaranteed

P. Wang and W. Zhuang, *Distributed Medium Access Control in Wireless Networks*,
SpringerBriefs in Computer Science, DOI 10.1007/978-1-4614-6602-4_7,
© The Author(s) 2013

priority access to voice traffic. The proposed scheme is distributed, collision-free, simple, and easy to implement. This work provides great flexibility and facility to the network service provider for service class management.

- In Chap. 5, we develop a dual busy-tone based MAC scheme for wireless ad hoc networks supporting voice and data traffic. The MAC layer QoS provisioning is more challenging in wireless ad hoc networks than in WLANs because the multi-hop network environment of wireless ad hoc networks raises a number of problems which do not occur in WLANs. In addition to the notorious hidden terminal and exposed terminal problems, the location-dependent contention may cause serious unfairness and priority reversal problems. All these problems can severely degrade network performance. Without solving these problems, QoS provisioning is difficult to achieve. Although some research work has been done, trying to address some of these problems, to the best of our knowledge, none of them provide a comprehensive solution to solve all these problems. The proposed scheme is the first one to utilize the busy-tones to address all these problems. As compared with the IEEE 802.11e, the proposed scheme significantly increases the system throughput by resolving the hidden and exposed terminal problems, greatly reduces voice traffic delay by ensuring guaranteed priority access for voice traffic, and considerably improves fairness performance for data traffic.

- In Chap. 6, we propose a distributed collision-free MAC scheme for a single-channel wireless mesh backbone to provide QoS support for multimedia applications. Different from the existing MAC schemes, our MAC scheme design benefits greatly from the fixed network topology of a wireless mesh backbone. With the router location information, collision-free transmissions are scheduled in a deterministic way, without the RTS/CTS handshaking prior to every packet transmission. Thus, the overhead is greatly reduced, as compared with contention-based MAC schemes. Meanwhile, the deterministic scheduling in our scheme is adaptive to the traffic dynamic and can achieve maximal spatial frequency reuse. By eliminating collisions, reducing overhead, and achieving maximal spatial reuse, the proposed scheme achieves much higher resource utilization than contention-based schemes. In addition, the proposed scheme can provide guaranteed priority access to real-time traffic and, at the same time, ensure fair channel access to data traffic. Unlike most of the existing MAC schemes which are limited to single-hop communications, the proposed MAC scheme takes the end-to-end QoS provisioning for multi-hop flows into consideration.

References

1. Wireless LAN Medium Access Control (MAC) and Physical Layer(PHY) Specifications, IEEE 802.11 standard (1999)
2. IEEE 802.11 WG, IEEE 802.11g, IEEE Standard for Information technology-Telecommunications and information exchange between systems-Local and metropolitan area networks-Specific requirements – Part 11: Wireless LAN Medium Access Control (MAC) and Physical Layer (PHY) specifications: Amendment 4: Further Higher Data Rate Extension in the 2.4 GHz Band (2003)
3. IEEE Standard for Information technology-Telecommunications and information exchange between systems-Local and metropolitan area networks-Specific requirements-Part 11: Wireless Medium Access Control (MAC) and Physical Layer (PHY) specifications: Amendment 7: Medium Access Control(MAC) Quality of Service (QoS) Enhancements. IEEE 802.11 WG, IEEE 802.11e/D11 (2004)
4. IEEE Standard for Local and metropolitan area networks Part 16: Air Interface for Fixed and Mobile Broadband Wireless Access Systems Amendment 2: Physical and Medium Access Control Layers for Combined Fixed and Mobile Operation in Licensed Bands and Corrigendum 1 (2006)
5. Aad, I., Castelluccia, C.: Differentiation mechanisms for IEEE 802.11. In: Proc. IEEE INFOCOM, vol. 1, pp. 209–218 (2001)
6. Ahn, C.W., Kang, C.G., Cho, Y.Z.: Soft reservation multiple access with priority assignment (SRMA/PA): a novel MAC protocol for QoS-guaranteed integrated services in mobile ad hoc networks. In: Proc. IEEE VTC, vol. 2, pp. 942–947 (2000)
7. Akyildiz, I.F., Wang, X., Wang, W.: Wireless mesh networks: a survey. Computer Networks **47**(4), 445–487 (2005)
8. Banchs, A., Perez, X.: Distributed weighted fair queuing in 802.11 wireless LAN. In: Proc. IEEE ICC, vol. 5, pp. 3121–3127 (2002)
9. Barry, M., Campbell, A.T., Veres, A.: Distributed control algorithms for service differentiation in wireless packet networks. In: Proc. IEEE INFOCOM, vol. 1, pp. 582–590 (2001)
10. Bharghavan, V., Demers, A., Shenker, S., Zhang, L.: MACAW: A media access protocol for wireless LAN's. In: Proc. ACM SIGCOMM, pp. 212–225 (1994)
11. Bianchi, G.: Performance analysis of the IEEE 802.11 distributed coordination function. IEEE J. Select. Areas Commun. **18**(3), 535–547 (2000)
12. Black, U.: Voice Over IP. Prentice Hall (2000)
13. Bladwin, R.O., IV, N.J.D., Midkiff, S.F., Raines, R.A.: Packetized voice transmission using RT-MAC, a wireless real-time medium access control protocol. Mobile Computing and Communications Review **5**(3), 11–25 (2001)

14. Bruno, R., Conti, M., Gregori, E.: Mesh networks: commodity multihop ad hoc networks. IEEE Commun. Mag. **43**(3), 123–134 (2005)
15. Cai, L., Shen, X., Mark, J.W., Cai, L., Xiao, Y.: Voice capacity analysis of WLAN with unbalanced traffic. IEEE Trans. Veh. Technol. **55**(3), 752–761 (2006)
16. Carlson, E., Prehofer, C., Bettstetter, C., Karl, H., Wolisz, A.: A distributed end-to-end reservation protocol for IEEE 802.11-based wireless mesh networks. IEEE J. Select. Areas Commun. **24**(11), 2018–2027 (2006)
17. Casner, S., Jacobson, V.: Compressing IP/UDP/RTP Headers for Low-Speed Serial Links. IETF RFC 2508 (1999)
18. Chen, D., Garg, S., Kappes, M., Trivedi, K.: Supporting VBR VoIP traffic in IEEE 802.11 WLAN in PCF mode. Avaya Laboratories (2002)
19. Chen, J., Chan, S.H., He, J., Liew, S.C.: Mixed-mode WLAN: the integration of ad hoc mode with wireless LAN infrastructure. In: Proc. IEEE GLOBECOM, pp. 231–235 (2003)
20. Cheng, H.T., Jiang, H., Zhuang, W.: Distributed medium access control for wireless mesh networks. Wireless Communications and Mobile Computing **6**(6), 845–864 (2006)
21. Crawley, E., Nair, R., Rajagopalan, B., Sandick, H.: A framework for QoS based routing in the Internet. Internet IETF RFC2386 (1998)
22. Deng, D.J., Chang, R.S.: A priority scheme for IEEE 802.11 DCF access method. IEICE Trans. Commun. **E82-B(1)**, 96–102 (1999)
23. Dovrolis, C., Ramanathan, P.: A case for relative differentiated services and the proportional differentiation model. IEEE Network **13**(5), 26–34 (1999)
24. Elson, J., Girod, L., Estrin, D.: Fine-grained network time synchronization using reference broadcasts. ACM SIGOPS Operating Systems Review **36**, 147–163 (2002)
25. Eshet, J., Liang, B.: The RRMS Protocol: Fair Medium Access in Ad Hoc Networks. In: Proc. of the 22nd Biennial Symposium on Communications, pp. 162–164 (2004)
26. Eshet, J., Liang, B.: Randomly Ranked Mini Slots for Fair and Efficient Medium Access Control in Ad Hoc Networks. IEEE Trans. Mobile Comput. **6**(5), 481–493 (2007)
27. Fan, W.F., Tsang, D., Bensaou, B.: Admission control for variable bit rate traffic using variable service interval in IEEE 802.11e WLANs. In: Proc. IEEE ICCCN, pp. 447–453 (2004)
28. Gambiroza, V., Sadeghi, B., Knightly, E.W.: End-to-end performance and fairness in multihop wireless backhaul networks. In: Proc. ACM/IEEE MOBICOM, pp. 287–301 (2004)
29. Ganeriwal, S., Kumar, R., Srivastava, M.: Timing-sync protocol for sensor networks. In: Proc. the first international conference on Embedded networked sensor systems, pp. 138–149 (2003)
30. Garg, S., Kappes, M.: Admission control for VoIP traffic in IEEE 802.11 networks. In: Proc. IEEE GLOBECOM, pp. 3514–3518 (2003)
31. Garg, S., Kappes, M.: An experimental study of throughput for UDP and VoIP traffic in IEEE 802.11b networks. In: Proc. IEEE WCNC, pp. 1748–1753 (2003)
32. Gummalla, A.C.V., Limb, J.O.: Design of an access mechanism for a high speed distributed wireless LAN. IEEE J. Select. Areas Commun. **18**(9), 1740–1750 (2000)
33. Haas, Z.J., Deng, J.: Dual busy tone multiple access (DBTMA) — a multiple access control scheme for ad hoc networks. IEEE Trans. Commun. **50**(6), 975–985 (2002)
34. Hastings, W.: Monte Carlo sampling methods using Markov chains and their applications. Biometrika **57**(1), 97–109 (1970)
35. He, Q., Cai, L., Shen, X., Ho, P.H.: Improving tcp performance over wireless ad hoc networks with busy tone assisted scheme. EURASIP Journal on Wireless Communications and Networking **2006** (2006)
36. Hiraguri, T., Ichikawa, T., Iizuka, M., Morikura, M.: Novel multiple access protocol for voice over IP in wireless LAN. In: Proc. IEEE ISCC, pp. 517–523 (2002)
37. Hole, D.P., Tobagi, F.A.: Capacity of an IEEE 802.11b wireless LAN supporting VoIP. In: Proc. IEEE ICC, pp. 196–201 (2004)
38. Huang, V., Zhuang, W.: QoS-oriented access control for 4G mobile multimedia CDMA communications. IEEE Commun. Mag. **40**(3), 118–125 (2002)

39. Huang, X.L., Bensaou, B.: On max-min fairness and scheduling in wireless ad-hoc networks: analytical framework and implementation. In: Proc. ACM MOBIHOC, pp. 221–231 (2001)

40. Jain, R., Durresi, A., Babic, G.: Throughput fairness index: an explanation. ATM Forum Document Number: ATM_Forum/99-0045 (1999)

41. Janssen, J., Vleeschauwer, D.D., Buchli, M., Petit, G.H.: Assessing voice quality in packet-based telephony. IEEE Internet Computing **6**, 48–56 (2002)

42. Jiang, H., Wang, P., Zhuang, W.: A distributed channel access scheme with guaranteed priority and enhanced fairness. IEEE Trans. Wireless Commun. **6**(6), 2114–2125 (2007)

43. Jiang, H., Wang, P., Zhuang, W., Shen, X.: An interference aware distributed resource management scheme for CDMA-based wireless mesh backbone. IEEE Trans. Wireless Commun. **6**(12), 4558–4567 (2007)

44. Jiang, H., Zhuang, W., Shen, X., Abdrabou, A., Wang, P.: Differentiated services for wireless mesh backbone. IEEE Wireless Commun. Mag. **44**(7), 113–119 (2006)

45. Jiang, S., Rao, J., He, D., Ling, X., Ko, C.: A simple distributed PRMA for MANETs. IEEE Trans. Veh. Technol. **51**(2), 293–305 (2002)

46. Jurdak, R., Lopes, C.V., Baldi, P.: A survey, classification and comparative analysis of medium access control protocols for ad hoc networks. IEEE Communications Surveys and Tutorials **6**(1), 2–16 (2004)

47. Karn, P.: MACA — a new channel access method for packet radio. In: Proc. ARRL/CRRL Amateur Radio 9th Computer Networking Conference, vol. 37, pp. 134–140 (1990)

48. Kim, H., Hou, J.C.: Improving protocol capacity for UDP/TCP traffic with model-based frame scheduling in IEEE 802.11-operated WLANs. IEEE J. Select. Areas Commun. **22**(10), 1987–2003 (2004)

49. Kleinrock, L.: Queueing Systems. vol 1, New York:Wiley (1975)

50. Kleinrock, L., Tobagi, F.A.: Packet switching in radio channels: part I — carrier sense multiple-access modes and their throughput-delay characteristics. IEEE Trans. Commun. **23**(12), 1400–1416 (1975)

51. Koksal, C.E., Kassab, H., Balakrishnan, H.: An analysis of short-term fairness in wireless media access protocols. In: Proc. ACM SIGMETRICS, pp. 118–119 (2000)

52. Kuo, Y.L., Lu, C.H., Wu, E., Chen, G.H.: An admission control strategy for differentiated services in IEEE 802.11. In: Proc. IEEE GLOBECOM, pp. 707–712 (2003)

53. Larzon, L.A., Hannu, H., Jonsson, L.E., Svanbro, K.: Efficient transport of voice over IP over cellular links. In: Proc. IEEE GLOBECOM, pp. 1669–1676 (2000)

54. Liang, B., Dong, M.: Balancing distance and lifetime in delay constrained ad hoc networks. In: Proc. ACM MOBIHOC, pp. 97–107 (2006)

55. Liang, B., Dong, M.: Packet Prioritization in Multihop Latency Aware Scheduling for Delay Constrained Communication. IEEE J. Select. Areas Commun. **25**(4), 819–830 (2007)

56. Lin, C.R., Geria, M.: Asynchronous multimedia multihop wireless network. In: Proc. IEEE INFOCOM, vol. 1, pp. 118–125 (1997)

57. Liu, H., Li, G.: OFDM-based Broadband Wireless Networks: Design and Optimization. Wiley-Interscience (2005)

58. Luo, H., Cheng, J., Lu, S.: Self-coordinating localized fair queueing in wireless ad hoc networks. IEEE Trans. Mobile Comput. **3**(1), 86–98 (2004)

59. Mahapatra, P., Li, J., Gui, C.: QoS in mobile ad hoc networks. IEEE Commun. Mag. **10**(3), 44–52 (2003)

60. Mangold, S., Choi, S., May, P., Klein, O., Hiertz, G., Stibor, L.: IEEE 802.11e wireless LAN for quality of service. In: Proc. European Wireless Conf., vol. 18, pp. 32–39 (2002)

61. Mark, J.W., Zhuang, W.: Wireless Communications and Networking. Prentice Hall, ISBN0-13-040905-7 (2003)

62. Metropolis, N., Rosenbluth, A., Rosenbluth, M., Teller, A., Teller, E.: Equations of state calculations by fast computing machines. J. Chem. Phys. **21**(6), 1087–1092 (1953)

63. Muir, A., Garcia-Luna-Aceves, J.J.: Group allocation multiple access in single-channel wireless LANs. In: Proc. Commun. Networks Distrib. Syst. Modeling and Simulation Conf. (1997)

64. Nandagopal, T., Kim, T., Gao, X., Bharghavan, V.: Achieving MAC layer fairness in wireless packet networks. In: Proc. ACM/IEEE MOBICOM, pp. 87–98 (2000)
65. N.Dimitriou, Tafazolli, R.: Quality of service for multimedia CDMA. IEEE Wireless Commun. Mag. **38**(7), 88–94 (2000)
66. Ni, Q., Romdhani, L., Turletti, T.: A survey of QoS enhancements for IEEE 802.11 wireless LAN. Wireless Communications and Mobile Computing **4**(5), 547–566 (2004)
67. Ong, E., Loke, M., Lin, W., Lu, Z., Yao, S.: Video quality metrics - an analysis for low bit rate videos. In: Proc. IEEE ICASSP, vol. 1, pp. I–889–I–892 (2007)
68. Pattara-atikom, W., Banerjee, S., Krishnamurthy, P.: Starvation prevention and quality of service in wireless LANs. In: Proc. IEEE WPMC, vol. 3, pp. 1078–1082 (2002)
69. Pickholtz, R.L., Milstein, L.B., Schilling, D.L.: Spread spectrum for mobile communications. IEEE Trans. Veh. Technol. **40**(2), 313–322 (1991)
70. Pong, D., Moors, T.: Call admission control for IEEE 802.11 contention access mechanism. In: Proc. IEEE GLOBECOM, pp. 174–178 (2003)
71. Robichaud, L.P.A., Boisvert, M., Robert, J.: Signal Flow Graphs and Applications. Prentice-Hall: Englewood Cliffs, New Jersey (1962)
72. Robinson, J.W., Randhawa, T.S.: Saturation throughput analysis of IEEE 802.11e enhanced distributed coordination function. IEEE J. Select. Areas Commun. **22**(5), 917–928 (2004)
73. Schwartz, M.: Broadband Integrated Networks. Prentice-Hall International, Inc (1996)
74. Sheu, S.T., Sheu, T.F.: DBASE: a distributed bandwidth allocation/sharing/extension protocol for multimedia over IEEE 802.11 ad hoc wireless LAN. In: Proc. IEEE INFOCOM, vol. 3, pp. 1558–1567 (2001)
75. Sobrinho, J.L., Krishnakumar, A.S.: Quality-of-service in ad hoc carrier sense multiple access wireless networks. IEEE J. Select. Areas Commun. **17**(8), 1353–1368 (1999)
76. Tabata, K., Kishi, Y., Konishi, S., Nomoto, S.: A study on the autonomous network synchronization scheme for mesh wireless network. In: Proc. IEEE PIMRC, vol. 1, pp. 829–833 (2003)
77. Tian, Z., Luo, X., Giannakis, G.B.: Cross-layer sensor network synchronization. In: Proc. the 38th Asilomar Conference on Signals, Systems and Computers, vol. 1 (2004)
78. Tobagi, F.A., Kleinrock, L.: Packet switching in radio channels: Part II—The hidden terminal problem in carrier sense multiple-access and the busy-tone solution. IEEE Trans. Commun. **23**(12), 1417–1433 (1975)
79. Toh, C.K.: Ad Hoc Mobile Wireless Networks: Protocols and Systems. Prentice Hall PTR (2001)
80. Vaidya, N.H., Bahl, P., Gupta, S.: Distributed fair scheduling in a wireless LAN. In: Proc. ACM/IEEE MOBICOM, pp. 167–178 (2000)
81. Veeraraghavan, M., Cocker, N., Moors, T.: Support of voice services in IEEE 802.11 wireless LANs. In: Proc. IEEE INFOCOM, pp. 488–497 (2001)
82. Wang, P., Jiang, H., Zhuang, W.: A dual busy-tone MAC scheme supporting voice/data traffic in wireless ad hoc networks. In: Proc. IEEE GLOBECOM, pp. 1–5 (2006)
83. Wang, P., Jiang, H., Zhuang, W.: IEEE 802.11e enhancement for voice service. IEEE Wireless Commun. Mag. **13**(1), 30–35 (2006)
84. Wang, P., Jiang, H., Zhuang, W.: Capacity improvement and analysis for voice/data traffic over WLANs. IEEE Trans. Wireless Commun. **6**(4), 1530–1541 (2007)
85. Wang, P., Jiang, H., Zhuang, W.: A new MAC scheme supporting voice/data traffic in wireless ad hoc networks. IEEE Trans. Mobile Comput. **7**(12), 1491–1503 (2008)
86. Wang, P., Zhuang, W.: A collision-free MAC scheme for multimedia wireless mesh backbone. submitted to IEEE Trans. Wireless Commun.
87. Wang, P., Zhuang, W.: An improved busy-tone solution for collision avoidance in mobile ad hoc networks. In: Proc. IEEE ICC, vol. 8, pp. 3802–3807 (2006)
88. Wang, P., Zhuang, W.: A token-based scheduling scheme for WLANs and its performance analysis. In: Proc. IEEE ICC, pp. 3716–3721 (2007)
89. Wang, P., Zhuang, W.: A collision-free MAC scheme for multimedia wireless mesh backbone. In: Proc. IEEE ICC (2008)

90. Wang, P., Zhuang, W.: A token-based scheduling scheme for WLANs supporting voice/data traffic and its performance analysis. IEEE Trans. Wireless Commun. **7**(5), 1708–1718 (2008)

91. Wang, W., Liew, S.C., Li, V.: Solutions to performance problems in VoIP over a 802.11 wireless LAN. IEEE Trans. Veh. Technol. **54**(1), 366–384 (2005)

92. Wang, Y., Garcia-Luna-Aceves, J.J.: Collision avoidance in multi-hop ad hoc networks. In: Proc. IEEE MASCOTS, pp. 145–154 (2002)

93. Wang, Y., Zhang, Y., Ostermann, J.: Video Processing and Communications. Prentice Hall PTR Upper Saddle River, NJ, USA (2001)

94. Wilson, J.M.: The next generation of wireless LAN emerges with 802.11n. Technology Intel Mag. (2004)

95. Wright, D.J.: Voice over Packet Networks. John Wiley and Sons (2001)

96. Xiao, Y.: IEEE 802.11n: enhancements for higher throughput in wireless LANs. IEEE Wireless Commun. Mag. **12**(6), 82–91 (2005)

97. Xiao, Y., Li, H., Choi, S.: Protection and guarantee for voice and video traffic in IEEE 802.11e wireless LANs. In: Proc. IEEE INFOCOM, pp. 2152–2162 (2004)

98. Yan, Y., Tomisawa, M., Gong, Y., Guan, Y., Wu, G., Law, C.: Joint timing and frequency synchronization for IEEE 802.16 OFDM systems. In: Proc. IEEE Mobile WiMAX Symposium, pp. 25–29 (2007)

99. Yang, X., Vaidya, N.: Priority scheduling in wireless ad hoc networks. In: Proc. ACM/IEEE MOBICOM, pp. 71–79 (2002)

100. You, T., Yeh, C.H., Hassanein, H.: CSMA/IC: A new class of collision-free MAC protocols for ad hoc wireless Networks. In: Proc. IEEE ISCC, pp. 843–848 (2003)

101. Zhai, H., Kwon, Y., Fang, Y.: Performance analysis of IEEE 802.11 MAC protocols in wireless LANs. Wireless Communications and Mobile Computing **4**(8), 917–931 (2004)

102. Zhai, H., Wang, J., Fang, Y., Wu, D.: A dual-channel MAC protocol for mobile ad hoc networks. In: Proc. IEEE GLOBECOM, pp. 27–32 (2004)

103. Zhao, R., Walke, B.: Decentrally controlled wireless multi-hop mesh networks for high quality multi-media communications. In: Proc. ACM MSWiM, pp. 200–208 (2005)

104. Zhu, H., Ming, L., Chlamtac, I., Prabhakaran, B.: A survey of quality of service in IEEE 802.11 networks. IEEE Commun. Mag. **11**(4), 6–14 (2004)

105. Ziouva, E., Antonakopoulos, T.: A dynamically adaptable polling scheme for voice support in IEEE802.11 networks. Computer Communications **26**(2), 129–142 (2003)